Money-Saving Strategies for the Owner/Builder

ROBERT L. ROY

author of "Underground Houses"

Sterling Publishing Co., Inc. New York

For Buz and Jean, Susan, Pat, Frank and Elizabeth, Ron and Debbie, and all the other heroes who have helped wear the path a little smoother for those that follow.

Special thanks to Jaki for her proofreading of the manuscript and her patience. And to Mark Stowe for the drawings on pages 103 and 106.

Other Books by Robert L. Roy
Cordwood Masonry Houses: A Practical Guide for the Owner-Builder
Underground Houses: How to Build a Low-Cost Home

Library of Congress Cataloging in Publication Data

Roy, Robert L.
　Money-saving strategies for the owner/builder.

　Includes index.
　1. House construction—Amateurs' manuals.
2. Dwellings—Energy conservation.　I. Title.
TH4815.R69　1981　　　690'.837　　　81-8843
ISBN 0-8069-7548-2 (pbk.)　　　　AACR2

Second Printing, 1982
Copyright © 1981 by Robert L. Roy
Two Park Avenue, New York, N.Y. 10016
Distributed in Australia by Oak Tree Press Co., Ltd.
P.O. Box J34, Brickfield Hill, Sydney 2000, N.S.W.
Distributed in the United Kingdom by Blandford Press
Link House, West Street, Poole, Dorset BH15 1LL, England
Distributed in Canada by Oak Tree Press Ltd.
% Canadian Manda Group, 215 Lakeshore Boulevard East
Toronto, Ontario M5A 3W9
Manufactured in the United States of America
All rights reserved
Library of Congress Catalog Card No.: 81-8843
Sterling ISBN 0-8069-7548-2

Contents

	Preface	5
1	Advantages of Owner-Building	7
2	The Land	12
3	The Grubstake	43
4	The Temporary Shelter	53
5	The Low-Cost Home	86
6	The People Who Did It!	123
7	Conclusions	146
	Appendix	148
	Annotated Bibliography	152
	Source Notes	159
	Index	160

Preface

Over the past six years, my wife Jaki and I have talked to perhaps thousands of people about building their own home. Hundreds visit our homestead each year and we meet many hundreds more at seminars, conferences, and energy fairs. Until now, I have concentrated on the actual construction techniques, especially in the fields of cordwood masonry-*Cordwood Masonry Houses: A Practical Guide for the Owner-Builder* (Sterling, 1980), and earth-sheltered housing-*Underground Houses: How to Build a Low-Cost Home* (Sterling, 1979). While these books touch on the economy of owner-building by accenting low-cost building techniques, they do not present a system of personal economic strategies for building your own home. This book should provide that "missing link."

Many—if not *most*—Americans accept and live by a prepackaged, standardized economic philosophy. The result is that they jump for joy when a bank "condescends" to loan them mortgage money on a 30-year note. They sign up for half a lifetime of fiscal servitude, beaming with their good "fortune." The radiance is short-lived, of course. Soon these beneficiaries of the bank's "benevolence" descend into that state which Thoreau so aptly calls "quiet desperation." And all this in the name of *security!*

I strongly urge you to read the article reprinted here with kind permission of the Plattsburgh Press-Republican.

Unfortunately, this mania for security is based on fear—fear of losing the "gold and silver fetters," as Thoreau so rightly calls our material values. As a result, we are insurance poor, terrified of sickness and death, crippled with building codes and planning restrictions, addicted to social security and pension schemes and so narrowly educated by the stalwart pillars of our society that what we learn is for *their* benefit, not ours.

The consequence of all this is that we accomplish exactly the opposite of what we set out to do. Rather than attaining greater security, we are very much less secure because *we delegate the responsibility for our own survival to others.*

The basic survival triumvirate—shelter, food and fuel—should be designed to work interdependently. *Shelter* should attend not only to itself, but should also address the "necessary of life," *food,* by aiding in its production and preservation. *Fuel* can be saved and stored by efficient construction and integration with locally available energy sources. In this way, we can become liberated from the long-term shelter costs of rent and mortgages.

You may be wondering about my own situation. Many people have asked me, "How were you able to finance the building of the Roy homestead?" My analysis is that (almost unconsciously) I had used certain economic strategies I'd absorbed through traveling and through reading, especially *Walden* and, perhaps surprisingly, *Siddhartha* by Hermann Hesse. The very way our community of owner-builders united and found a piece of land in northern New York generated other economic advantages, such as collective land buying and labor bartering. Thus, our good fortune was a combination of circumstances and, for my part, the illuminating experience of spending the eight or ten most formative years of my adult life outside the confines of these borders. During that time, I experimented with one of the strategies offered in this book—renovation—although I did not see it as a "strategy" at the time. Later, the accumulated value of renovating the old, stone, Scottish farm cottage in which we lived was converted to the grubstake which financed our American homesteading adventure. So, without knowing it, we had adopted yet another strategy for remaining fiscally free at least.

Others in our hilltop community arrived at similar results, but by different routes, and they have generously consented to share their stories and strategies.

This is a new kind of book, a kind of do-it-yourself brand of home financing. But, before launching into this treatise, it should be emphasized that there exists a series of paradoxes regarding low-cost owner-building. One such dilemma is that the ever more stringent building and planning codes are pushing people who want to build their own homes, especially homes that show any imagination or which use natural (as opposed to *manufactured*) building materials, further and further away from the population centers even though the code restrictions were originally created to protect tenants and home buyers from unscrupulous landlords and contractors, not to restrict owner-building.

On the positive side, intelligent settlement on the edges of the wild has a great potential for discouraging thoughtless, wasteful development of the land for short-term financial gain. But this presents another problem. Preserving the wilderness requires involvement in local decision-making, an unappealing obligation to some people who move to the very fringes of human settlement specifically to escape the dictates of land use.

These are the contradictions and riddles that must be wrestled with in the near future. While I am steadfastly committed to a human being's right to provide his or her own shelter, the trend toward moving out to the wild must be critically examined. We may have already reached—or overextended—the safe limits of "taming" the wild. When the wild is gone, so, too, are all hopes of attaining true civilization, and planetary survival itself will be in doubt.

Press-Republican—Wednesday, November 5, 1980

Purchase of home remains tough for first-timers

Business News

Wall Street Journal-ONS

Most people want to own their own homes. But few can afford to buy that first one. "The first-time home buyer is the real loser today," says Sid Green, a real estate agent in Manchester, Conn. "With interest rates so high, and houses so expensive, these kids are really getting squeezed. How can they save as much as they need?"

They can't and so many have stopped looking for homes. First-time home buyers accounted for only 18 percent of the market last year, down from 36 percent two years earlier, according to a survey by the U.S. League of Savings Associations. Those who buy often do so only by borrowing from relatives and forgoing many of life's luxuries. A few are getting help from new types of mortgages and relaxed lending standards.

The numbers are striking. Take a $70,000 house, which is just a shade above the average price for an existing home. With a 20 percent down payment and a 30-year loan at 13½ percent interest rate, annual mortgage payments come to about $7,700. In 1977, the average existing home cost $47,500. With a similar loan but at the then prevailing 9 percent rate, annual mortgage payments were about $3,670.

"It's ridiculous. We can't afford to make mortgage payments that are double what we're paying for rent," says Georgia Keiss, 31 years old, who lives with her husband and two children in a two-bedroom apartment in Evanston, Ill. "We'd love to have a home. We're so tight in this place we've got kids coming out of our ears. But if we bought a house, we wouldn't eat for a year."

Still, the desire to own is strong. A 1978 Louis Harris poll showed that 93 percent of the prime first-time home buying group, those aged 25 to 34, wanted their own home. "I think it has surprised everyone just how much people have been willing to sacrifice to get that first home," says Bernard Frieden, professor of urban studies and planning at Massachusetts Institute of Technology.

Part of that sacrifice, Frieden says, is manifested in changing lifestyles. Couples are having children later, allowing both partners to work longer. They are returning to the cities, he says, "not because they want to live in cities, but because it's a first step on the ladder of home ownership. City homes are cheaper."

Furthermore, says Robert Sheehan, an economist with the National Association of Home Builders, first-time home buyers are going after more modest homes. He says that the average square footage of living space in new homes fell from 1,704 in 1979 third-quarter sales to 1,688 in the fourth quarter to 1,667 in this year's first quarter, the latest data available.

Even if the desire and the willingness to sacrifice are there, many people still can't get that first house because they don't meet lending standards. The traditional standard is that housing payments can't exceed 25 percent of the borrower's gross income. On a typical $56,000 mortgage at 13½ percent for instance, the borrower must make more than $35,589 a year to qualify.

But that rule may be changing. MGIC Investment Corp., the nation's largest private insurer of home mortgages, last month raised that level to 33 percent. In the example above, that would mean a person could make $26,958 a year and still qualify.

Although most people agree that the relaxed standards will bring in more home-seekers, some experts doubt that the effect will be significant. "It's just a gimmick," says Roland Barstow, chaiman of Bell Federal Savings & Loan Association in Chicago. "We aren't turning people down because they don't qualify. People aren't coming to us because they know they can't afford a house."

Frieden of MIT further warns that some of the current strategies to buy the first house may backfire. In particular, he says that the two-income method may fall prey to the high divorce rate and periods of high unemployment. Besides, he adds, "prices are getting so high that husbands may need to go to polygamy to buy a house. They'll need two working wives."

---

* AUTHOR'S NOTE: This comes to a total outlay of $245,000 for the $70,000 house.

1

Advantages of Owner-Building

The very act of building your own shelter will virtually guarantee you a substantial dollar savings because of the elimination of labor costs, profit and a proportional amount of bank interest, but even if this were not so, owner-building still would have many positive attractions, both practical and spiritual.

QUALITY

One of the most common fears of potential owner/builders is that they do not have the skills or knowledge to build a house with the high standards of the contractor-built home. This fear stems from the myth that building a home is a very complicated procedure. A great deal of experience is deemed mandatory, ergo, only a professional builder should be entrusted with the task. This line of uninformed reasoning can lead to ignorant conclusions, such as, "Owner/builders construct shacks."

The truth is that the quality of a home does not depend on whether or not it is owner-built or contractor-built. Quality depends, first and foremost, on the conscientious attention to detail. Thus, in many cases, the owner-built home is of higher quality because the owner's long-term interests are the prime motivating force rather than merely a profit or a wage reward for punching the clock. Now, this is not to say that professional builders are not craftsmen; many most certainly are. But craftsmanship can be present in the owner-built home, too.

While it is certainly true that most houses in America today are contractor-built (about 80 percent in 1980), it is also true that a large part of the world's population lives in "homemade" houses, many of which have harmonized

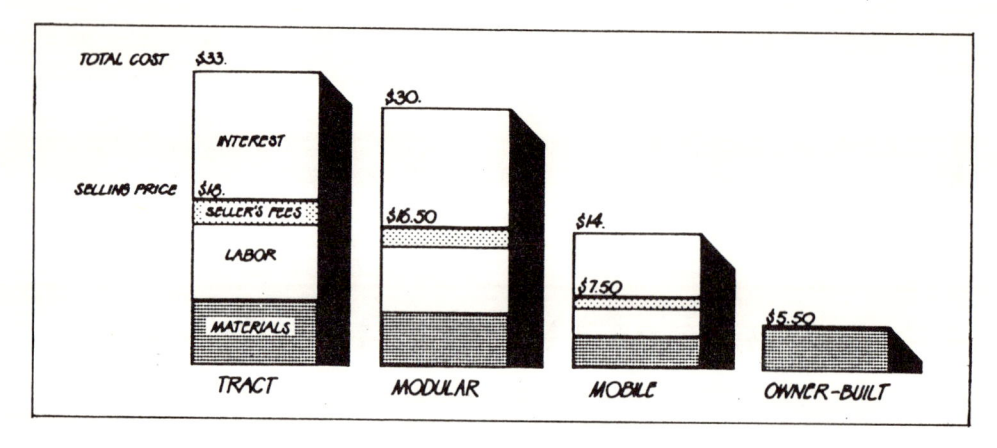

Figure 1. Here you can compare the per-square-foot cost of each of the least expensive types of housing with owner-built housing of comparable caliber. Interest is based on a 20-year loan at 7 percent. Keep in mind that the actual figures, which will vary, are less important than the relative overall values.

with the landscape for hundreds of years and more. (Consider, too, that contractors are unknown in the animal world where some excellent examples of craftsmanship can be found.) The truth is that building a home is not all that complicated. Owner/builder advocate Ken Kern says,

> "Not everyone has the attitude and physical ability necessary to begin a house and proceed to a successful conclusion. But most people, once they have made the initial decision to proceed, do find these qualities in themselves. Success is more a matter of determination than of previous experience."[1]

A few years ago, I was working with a construction crew on a house for an out-of-state customer. One morning, the contractor told us, "Look, the loan officer is coming to inspect the foundation at one o'clock. We're backfilling this afternoon. We've got to have the Thoroseal®"—a masonry sealer—"on the wall before we can backfill. Now, let's *go!*"

So we applied a single coat of Thoroseal® to the foundation, although a double coat is called for by the manufacturer *with a 24-hour wait between coats*. The bank loan officer was on time and saw that the Thoroseal®—still damp—was on the basement wall. He was satisfied. That afternoon, the foundation was backfilled. The bank was happy. The contractor saved a day. But, what did the customer get? A few days later, after another incident involving the structural integrity of a different house, I quit.

Previous to these events, my wife, Jaki, and I built our first home, Log End Cottage, and applied the Thoroseal® correctly to the block wall. We were inexperienced, but all we had to do was to follow the directions on the bag.

There are many excellent contractors, of course, who endeavor to put out a quality product, but they can't always be present to supervise every construction detail. From my experience on construction crews, I've learned that the workers do not always do things quite as carefully as they would on a job for themselves. Personal involvement on a project can provide that all-important difference.

You will make mistakes, to be sure, and there are certain skills—stone masonry, for example—that take years to master insofar as aesthetics are concerned. But a strong, long-lasting house can be built—even of stone—although it may not have quite the polished appearance found in the work of master masons.

EASE OF REPAIR AND MAINTENANCE

If something goes wrong in your contractor-built home, you would, in all likelihood, call in a contractor or tradesperson to repair it—especially if it's within the period of time (usually one or two years) during which the contractor is legally responsible. An owner/builder, on the other hand, would find it counterproductive to call in a tradesperson for normal maintenance or repairs. For a start, the tradesperson does not know the house at all, while the owner/builder is familiar with each nail and board. Having built his or her own home, owner/builders are rarely inclined to pay 15 or 20 dollars per hour to have someone else do the repairs. Therefore, the life-cycle cost—or "on-cost"—of the owner-built home is less.

COMFORT

Your home should not only be *built* by you, its future occupant, but *designed* by you as well. Can your architect know you better than you know yourself? As often as not, architects design homes at least partially to satisfy some personal, structural or aesthetic whim, and the client's requirements are moulded to that whim, instead of vice versa. Again, to quote Ken Kern:

> [Owner/builders] imagine being able to have a house of any shape they want—designed by themselves to meet their most practical needs and their most whimsical fancies. They wonder what it would be like if no one else made these decisions for them. What would it be like to be an artist house-builder in the only true sense—in a way that architects, who interpret clients' visions, and builders, who are allowed no visions at all, cannot? What would it be like to touch all the materials, to learn about placing them one against the other? What would it be like if the mistakes were made by their own hands instead of by the mechanisms of technolgy? What would it be like to have stories to tell about the creations of their houses?"[2]

An owner-built home fits the designer-occupant like an old shoe. A special comfort, as well as economic advantage, can be derived from using indigenous materials in the construction. A house of stone on a stony site will be in harmony with its surroundings, as will be a log house in a wooded site.

Owner-built homes usually are constructed on a pay-as-you-go basis. This is a sound and time-proven strategy, as we will see. One of the by-products of this building strategy, however, is that the original plans tend to change as time goes on. But change should be seen as a positive development when it's based on your *newest* requirements instead of your distantly anticipated requirements. In this way, the house moulds naturally to the "contours" of its inhabitants.

A change of plans in a contractor-built home normally carries a heavy penalty, and rightly so; it adversely affects how a contractor budgets time for his other work. Changes in code-approved plans involve unacceptable delays within the bureaucratic machine.

Special design features found in owner-built homes are often spontaneous ideas, the use of a burled branch as a handrail, for example, or bottles included in cordwood masonry as a purely aesthetic effect. It is a rare carpenter or mason who will take it upon himself to improvise. And, even if he did, *and* his improvisation was successful visually or functionally, it would still have been *his* idea. To the occupant, the meaning and feeling derived from the improvisation is of a different kind to that felt by the owner/builder, which leads us to:

SATISFACTION AND PRIDE

There is an intangible part of comfort which comes not so much from exactly what has been built, but rather *who built it*. Now this quality may only be manifest to the builder himself, and perhaps to his or her family and close friends, and would be largely lost if the house were sold to a stranger, but it is present as long as the builder occupies his own home.

Owner/builder Cliff Shockey, of Vanscoy, Saskatchewan, says, "The satisfaction of designing and building your own home cannot be measured. It is truly one of the most rewarding things I have done."[3]

Sam Felts says of his experience building a round cordwood house in Adel, Georgia, "This has been the most rewarding experience in my life. I feel that I have created a fun place in which to live."[4]

Bonny Pond, of Petit Rocher, New Brunswick, says, "Our house is more than a beautiful, ecologically sound, economically attuned domicile. It is a statement of what we feel about ourselves, each other and the world in which we live."[5]

An overly dramatic exaggeration? Not at all. And if, at times, owner/builders seem just that wee bit sinful in their pride, they are easily forgiven. They deserve it.

ECONOMICS

A little more than one-third of after-tax income is commonly spent on shelter, whether the shelter is rented or mortgaged. If an individual works from the age of 20 until 65, it can be argued that 15 years of his 45-year working life have been devoted to keeping a roof over his head. By the mortgage route, the individual has at least built up an equity in his home—albeit at a very high cost—and in most cases, the retirement years are relatively free of shelter costs, property taxes and normal maintenance excluded.

The rental route is an economic disaster in all but the most unusual individual circumstances, as no equity is built up at all. The mortgage road seems the better of the more common approaches toward paying for one's shelter. But, to obtain a mortgage commonly requires a substantial down payment which may even exceed the total land-and-materials cost of the owner-built home.

In northern New York in 1980, down payments on contractor-built homes reached as high as 30 percent, or $18,000 on the typical $60,000 house. Owner/builders in the same area were able to *buy their land, build their houses and pay for their living costs for less than the down payment on the mortgaged structure.*

Obviously, the magic key is the addition of the owner's labor. In some cases, people take six months off from regular employment to build, in others the construction is done after work and on weekends. People say to me, "That's okay for golf pros, writers and other layabouts, but I can't afford to take six months off work to build my house." I argue that to save the equivalent of 14½ years of their working life, they can't afford not to.

Obviously, building your own home does not guarantee retirement at 50 instead of 65, but the economic impact throughout the working years—however many they may be—is very much more favorable. The need to come up with a $400 mortgage payment each month for example, is a constant pressure—sometimes for 20, 30, years, or more. Someone else, usually a lending institution, literally owns a piece of the property. The loss of your job or any other circumstance making it difficult to come up with the payment can be disastrous, especially during the early years of the mortgage when most of the payment goes toward interest and little toward principal. Equity is minimal, so if you are forced to sell at this point you're virtually guaranteed a poor price.

There is another consideration with regard to making a mortgage payment. To put aside $400 to meet a payment requires earning considerably more than that figure. The almost $5,000 required over a year forces the occupant (we can't truly call you the "owner") to dwell in the world of the higher tax bracket. Depending on individual circumstances, the income taxes and social security taxes on the extra money needed for shelter might add from $1,000 to $2,000 to required earnings in this example. And, while a $400 monthly mortgage payment is not uncommon in our area because of higher down payment requirements and generally lower housing costs, the national average for new mortgages is a payment of $641 per month.

Through owner-building, less of your time need be spent on that otherwise expensive necessity of life called shelter. For many owner/builders, this means more free time. Or it can mean more money to spend on nonshelter-related items. Roy's First Law of Empiric Economics might be stated thus: Work to *save* money. Don't earn money to pay someone else to do what you can do yourself. A dollar saved is worth a whole lot more than a dollar earned, because we have to earn so darned many of them to save so precious few.

2

The Land

The cost of land may be the single greatest expense you'll incur in your owner-built home, so we will spend some time on the various economic considerations involved.

Where to Live

The first question you must ask yourself prior to the land search is, "In what sort of area do I want (or need) to live?" Central city? Suburbs? Exurbs (counties bordering metropolitan areas)? Countryside? Wilderness fringe? Consider future strategies as well as present factors. It may be that the strategy of leaving your job to build a house (discussed later) promises such compelling economic benefits that it won't be necessary to limit your land search to areas within commuting distance of your present job. On the other hand, while the economic and other personal benefits that come from living in your own house may be important to you—they may not be important enough to sacrifice a particularly satisfying position. Your job may tie you to a particular area, and the possible extra cost of the land, if located conveniently, may be a favorable trade-off, particularly if your salary is satisfactory.

However, low-cost owner-built homes generally are situated away from the population centers for three reasons. One, land prices near town are often prohibitive when compared to the house costs. Two, stricter planning regulations and building codes control these areas, adversely affecting some of the owner/builder's most valuable building strategies, such as the temporary shelter and the pay-as-you-go house. Three, other personal values, such as clean air and water, economic independence in food and fuel production, and peace and quiet are almost always important considerations to a person who makes the philosophical decision to be responsible for his or her own shelter . . . and life.

COUNTRY LAND IS CHEAPER, BUT . . .

Consider the targetlike illustration of Figure 2. The "bull's-eye," Area A, is a circle of 10 miles in diameter, representing all the land within 5 miles of downtown Centerville, a town of, say, 40,000 people. A total area of 78 square miles (202.02 sq km), it has a population density of 512 people per square mile (2.59 sq km). Even within this area, of course, the density varies. A residential neighborhood near the center might have 3,000 people in a given square mile. In cities, the density would even be higher. Nevertheless, most of the businesses, schools, churches and recreational facilities are found within the "bull's-eye" and all the people in this area are able to share the benefits of relatively close proximity to these amenities.

Here, land is sold in terms of lots and cannot even be considered on a per-acre basis. Owner/builders are limited by building and zoning regulations to constructing pretty much the same kind of home sold by contractors and developers.

Even here, however, there is a great potential for savings, as only about 40 percent of the selling cost of a contractor-built home goes towards materials. The lot price must be added, and the price *will* be higher than the average cost of a lot if a contractor is subdividing.

Then where are the potential savings? In labor costs, contractor and developer profit and a fair amount of bank interest. The problems you'll encounter are: bureaucratic hassles if you build anything even slightly different from "conventional standards"; severe restrictions on the add-on or pay-as-you-go strategies (temporary shelters are o-u-t!); high interim shelter costs during construction; and financing. More materials are generally needed, and more *expensive* materials at that, to meet town zoning and building code requirements. Even though a potential savings of 50 percent can be demonstrated, and a smaller loan is thus required, lending institutions will be reluctant to make mortgage money available to "inexperienced do-it-yourselfers." (A personal loan may be possible based on the strength of the borrower's reputation, but the interest will be much higher.)

While I believe that building your own home under these conditions is preferable to the lifetime rent or mortgage situation, especially if you're happy with your position and life in the town, it is not one of the strategies with which this book is primarily concerned. Complete or near-complete freedom from the shelter aspect of fiscal bondage requires an almost quantum leap away from the idea of an owner-built but conventional house in a subdivision.

My last comments on this are directed toward people who want the typical suburban ranch, don't want to build it themselves and think that the use of "alternative" materials and building styles is best left to hippies and beavers.

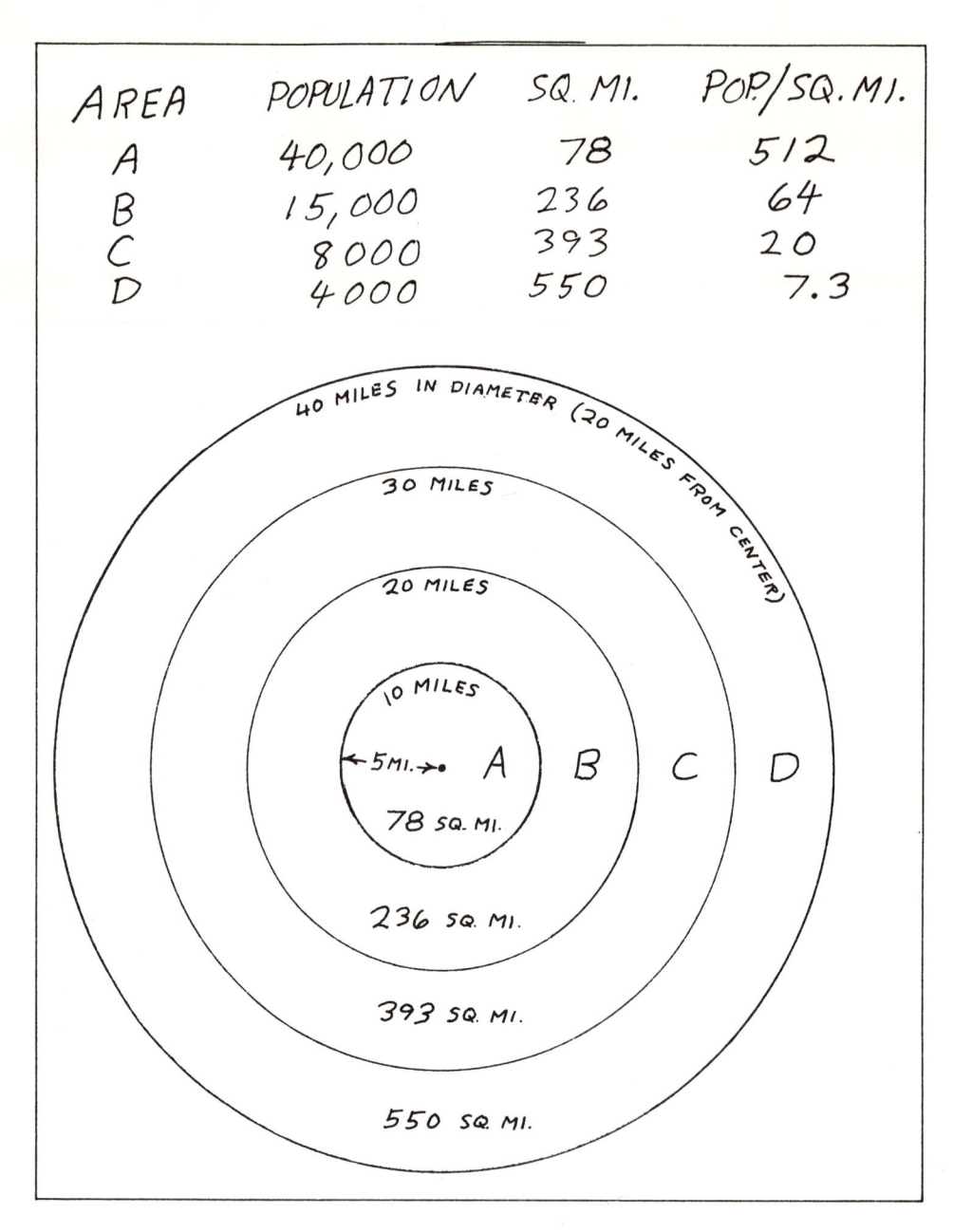

AREA	POPULATION	SQ. MI.	POP./SQ. MI.
A	40,000	78	512
B	15,000	236	64
C	8000	393	20
D	4000	550	7.3

(Diagram of four concentric circles labeled A, B, C, D — 40 miles in diameter (20 miles from center), 30 miles, 20 miles, 10 miles, with 5 mi. radius for A; areas: 78 sq. mi., 236 sq. mi., 393 sq. mi., 550 sq. mi.)

Figure 2. Above, the entire—but hypothetical—region known as "Centerville."

Frankly, I don't know how you've made it this far into the book, but you, too, "deserve a break today!" My advice is that you consider acting as your own general contractor. This does require some organization and research on your part, but 10 to 20 percent of the final cost can be saved in this way. Further savings are possible by keeping the house shape and structure simple, as outlined in Chapter 5. Otherwise, I can only invite you to read on. You may never put these strategies and philosophies to use, but then again, they might get you to thinking about alternatives.

The second ring, Area B in Figure 2, is the rest of the land within 10 miles of downtown Centerville not included in Area A. Its area is 236 square miles and might typically provide habitat for 15,000 more people. In a community like Centerville, these people often will be a combination of small farmers and overflow from the town. There will be individual home lots, mobile homes and even some planned housing developments. The population density drops to 64 per square mile (25 per sq km), which, incidentally, is just about the average for the United States as a whole.

Planning and zoning restrictions are not likely to be as severe as in Area A, although building codes may be, depending on where actual town boundaries fall and whether or not state building codes are in effect. (Keep in mind that this belt will be very much in demand for all kinds of development, as Area A is already nearly saturated and Area B is still within comfortable commuting distance.)

Proximity to town makes the prospect of settling here attractive to the owner/builder, too, although land prices are likely to be quite high because of the pressure from the expanding town. Even though the density is only one-eighth as great, building lots are inclined to be much bigger and much of the land is likely to be used in agriculture, so the net effect still is that of a lot of people chasing comparatively little "spare" land.

Area C is very popular among owner/builders. Population density is down to 20 per square mile (roughly 8,000 people spread over 393 square miles), and the greater distance from town eliminates much of the buying pressure on land prices. The proximity to employment opportunities is acceptable, although an assessment of the economic and time trade-offs involved in commuting might be necessary at this point. An economy car usually is the solution for owner/builders in this area.*

If a second vehicle is owned, often it is a small pick-up truck, useful for homesteading, building and other country activities. For me and my family,

---

* AUTHOR'S NOTE: An alternative strategy which merits consideration is often employed, especially by those starting out with a small grubstake (see Chapter 3). It is to buy a ten-year-old big American "gas hog" for about $200 and to run it into the ground. At first, this may seem wasteful and distasteful, but as far as energy consumption is concerned, the poor mileage must be counterbalanced with the energy cost involved in the manufacture of a replacement car. Many friends speak highly of the economics of this strategy. Reliability, however, is another matter. If you know instinctively that this is the strategy for you (or you wonder if it might be), you'll benefit from reading *Drive It 'Till It Drops,* by Joe Troise (And Books, The Distributors, 702 South Michigan, South Bend, IN. 46618).

a Toyota pick-up truck served all our transportation needs for two years and now, six years old, it is still serving us as we build our third house.

Area C is commonly devoted to agricultural use and, if it is prime farmland, it will not be cheap. In hilly or wooded areas, however, the price of land may be very favorable. The Roy homestead is located in just such country. Other, smaller towns often are found in this belt. When this is so, the population density for the area as a whole is higher than indicated, but not in the spaces between the villages.

Area D includes all the land between 15 and 20 miles from the center of Centerville. Population density is down to 7.3 people per square mile in this hypothetical model. Land will be quite a bit cheaper, barring special features such as lakefront, village land, and so on. Roads do not run "as the crow flies," however, and commuting to a regular job in Centerville might be unacceptable in terms of time and money. If you wanted to maintain a job in town, you'd be advised to weigh up this trade-off very carefully indeed, especially in light of the new reality of energy costs. A car pool with neighbors would be a financial aid, but the travel time still would be considerable.

Very often, owner-builders this far from town sever themselves from urban employment. They find work in the country closer to home, or they adopt a life-style in which they are able to support themselves with a home-centered business or part-time work. Here are to be found true homesteaders and farmers, living on low budgets by providing all their own firewood fuel and a significant amount of their food via their gardens and livestock. Usually, they pay less for property taxes, as well.

If there are building codes here—and very often there aren't—building inspectors are likely to be very flexible. But don't take this for granted. Try to find out from others in the area how flexible the building inspector really is.

Other important considerations in this area are: proximity to schools, doctors, and fire departments; snowplow and mail service; cost of power and telephone; nearest neighbors. Alternative energy sources, such as wind, solar and water power, all become potentially more economic. Clustering of like-minded families is a good strategy, one that will be illustrated in detail in other parts of this book.

Beyond the 20-mile ring—remember, this is a much greater distance from town if *road miles* are considered—the land may be cheap, but life is spartan and lacking in human companionship. Commuting is impractical, especially in the wintertime. Building here on the edge of the wilderness (*social,* if not *actual,* wilderness) requires a certain pioneer spirit (possessed by the very few), and an already established independence of the need for human companionship. Living in such a rural environment is both good for the kids and very hard on them. They will be better educated in nature and the fundamental realities of survival, but may be somewhat lacking in the humanities, even if they watch television on their 12-volt set. (*Especially* if they should!)

In this region, and perhaps in parts of Areas C and D, another responsibility confronts the homesteader (for he is now more truly a homesteader than simply an owner/builder). It is the responsibility to husband the land

and protect the existing vegetation, both flora and fauna. Ralph Borsodi (1887–1977), American economist and exponent of self-reliance, put it this way,[6]

> *BORSODI: I very carefully divide the possessions of mankind into two categories: one I call "property" and the other "trusterty." Now property, by definition, is anything which can be owned . . . legally owned. But you know there are some things that can be legally—but not morally—owned. For instance, slaves used to be legally owned. The statutes of our states and the Constitution of the United States made it legal to own human beings . . . but no amount of legalizing made it moral.*
>
> *I feel the same way about the natural resources of the earth. When you make something with your own labor you have, so to speak, frozen your labor into that thing. This is the way in which you create a moral title to that thing, by producing it. You can sell it to somebody else and, in return for what he pays you, you can give him your moral title to whatever it is.*
>
> *But no man created the earth or its natural resources. And no man or government has a moral title to the earth's ownership. If it is to be used, and we have to use it in order to live, then it has to be treated as a trust. We have to hold the earth in trust.*
>
> *We can enjoy the fruit of the land or of a natural resource, but the land or the resource itself must be treated as a gift. A man who uses the land is a trustee of that land and he must take care of it so that future generations will find it just as good, just as rich, as when he took possession of it. A trustee is entitled to a return for administering his trust . . . but he must never destroy the trust itself.*

The example of Figure 2 is greatly oversimplified, to be sure. Population centers larger or smaller than the fictitious Centerville will affect the sizes of the circles proportionally with regard to changes in land prices. The point is that the price of land reduces as the distance from town increases, other considerations being equal. Of course, other considerations are not equal. The type and quality of land is important, as well as such factors as water and power availability. These factors will be discussed.

If land were all the same, an equal concentric development around a centerpoint might well occur, as shown in Figure 2. In the real world, however, topography warps the theoretical development to something more closely resembling Figures 3a–3d. These illustrations show that smaller towns and settlements, satellites to the main town in the area, also exert an influence on land prices, as does the proximity of a more distant city. (Figure 3b) Grand City, with one million people, is assumed to be 75 miles from Centerville. This not-too-distant population exerts an influence on property values in the Centerville area, too, because of the desire of many Grand City people to own a vacation cabin or retirement home in the country.

The denser the cross-hatching in Figure 3c, the greater is the demand for the land. Of course, quality of access is just as important as distance. If an

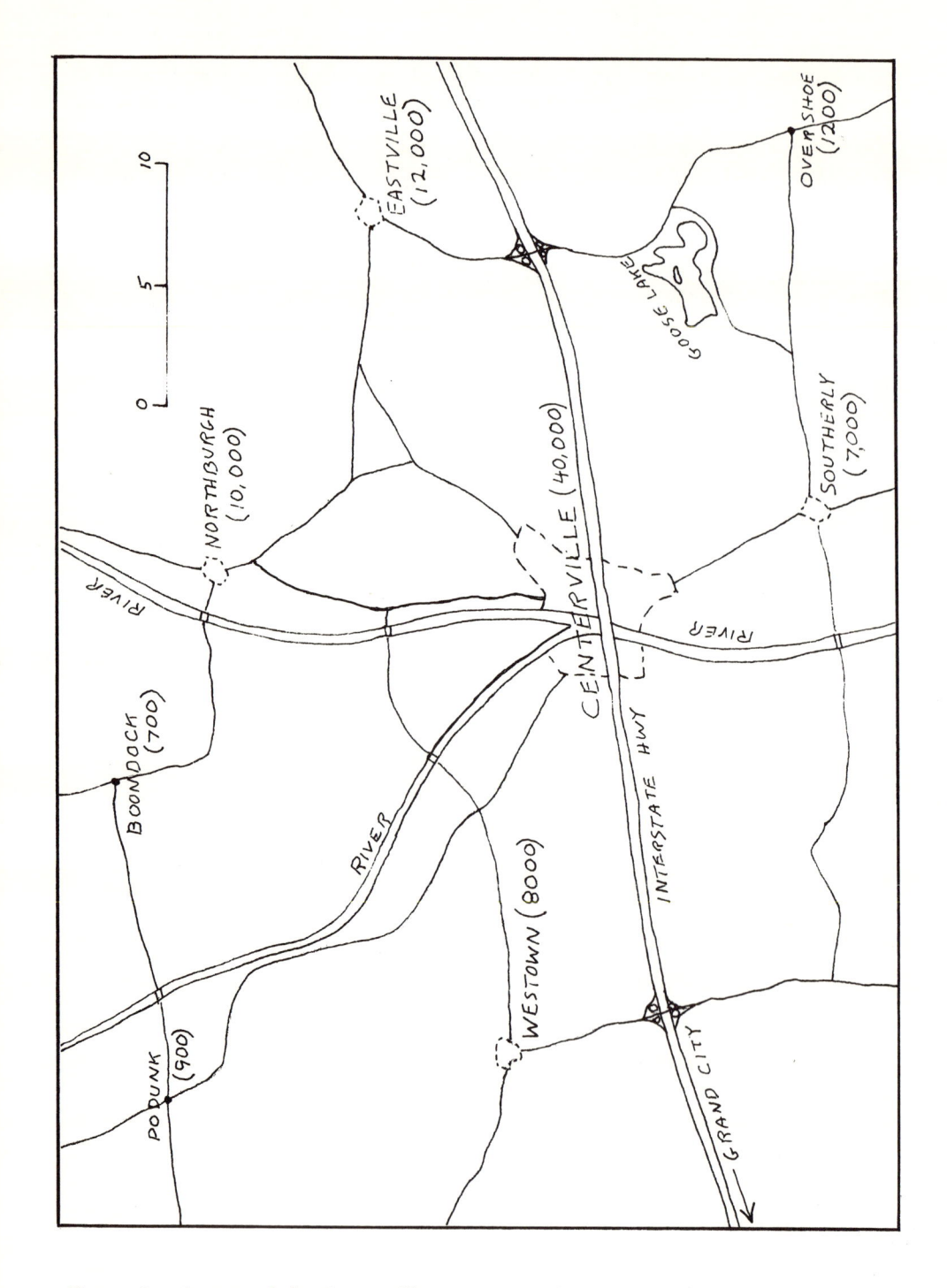

Figure 3a. A map of the Centerville area. Grand City is assumed to be 75 miles west of Centerville. All population estimates are in parentheses.

interstate highway exists between Centerville and Grand City, the "gravitational influence" is greater, as shown in Figure 3d. Similarly, properties on secondary roads and dirt tracks will be proportionally lower in cost than properties on state and county roads. Taken to an extreme, land without access at all can be bought very cheaply indeed, but of what use is it to the owner-builder if it can only be reached by parachute?

Like Figure 2, Figures 3a–3d are greatly simplified in that they do not take into consideration landform, quality of schools, quality of the existing development, details of access and utility availability. Generally, land quality and topography will influence the degree of effort to supply access. The resulting development and its proximity to various population bases will dictate the demand for, and cost of, the land. My examples are meant to illustrate some of the influences on land prices in an area developed to the extent suggested by the population figures for a place like Centerville and its surrounding towns. (See Figure 3a.)

Building and Planning Regulations

The original intent of the building codes was an honorable attempt to prevent and prohibit the very unhealthy living conditions imposed upon people by unscrupulous builders and owners of shoddy tenement houses. According to Rudolph Miller, founder of the Building Officials Conference of America (BOCA) in 1915, the purpose of the building code is as follows:

> The building laws should provide only for such requirements with respect to building construction and closely related matters as are absolutely necessary for the protection of persons who have no voice in the manner of construction or the arrangement of buildings with which they involuntarily come in contact. Thus, when buildings are comparatively small, are far apart, and their use is limited to the owners and builders of them, so that, in case of failures of any kind that are not a source of danger to others, no necessity for building restriction would exist.[7]

Today, the codes have evolved to the point where they present a very real obstacle to the owner/builder, financial and otherwise.

Planning (zoning) regulations are meant to protect inhabitants of a residential neighborhood from a factory moving within their midst, and other similar kinds of problems. The evolution of thinking with regard to planning has taken the emphasis off the protection of people's quality of life and on to the protection of "property values."* (Cont'd. on page 23.)

---

* AUTHOR'S NOTE: There are rays of sun on the planning landscape. The city of Davis, California, has pioneered building codes and regulations that, among other measures, require new houses to have a southerly orientation. Good planning is better than bad planning, and, in some cases, might be better than no planning, although I say this with caution.

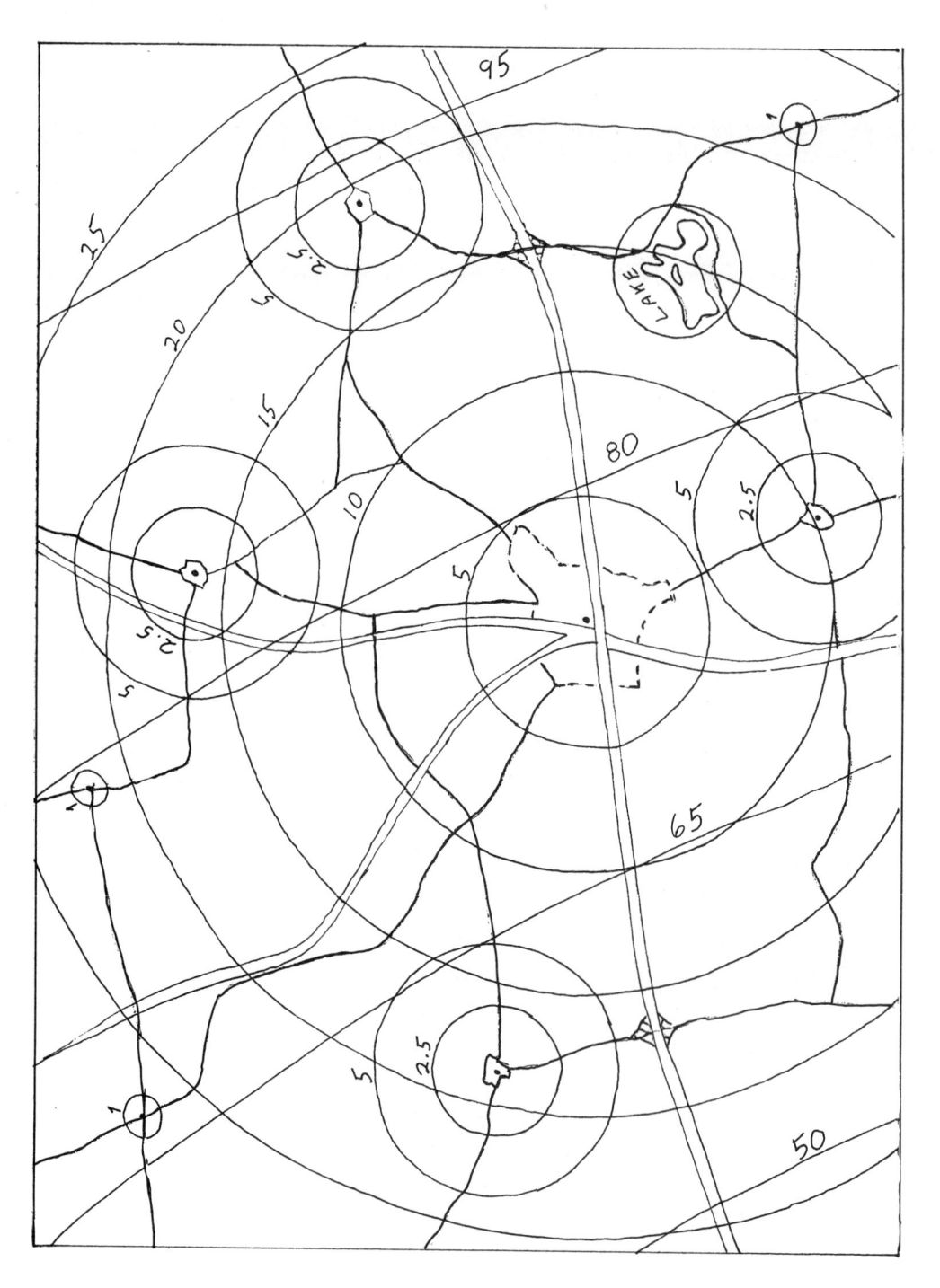

Figure 3b. In this diagram the spheres of influence on land values are delineated. The numbers signify approximate distances to population centers.

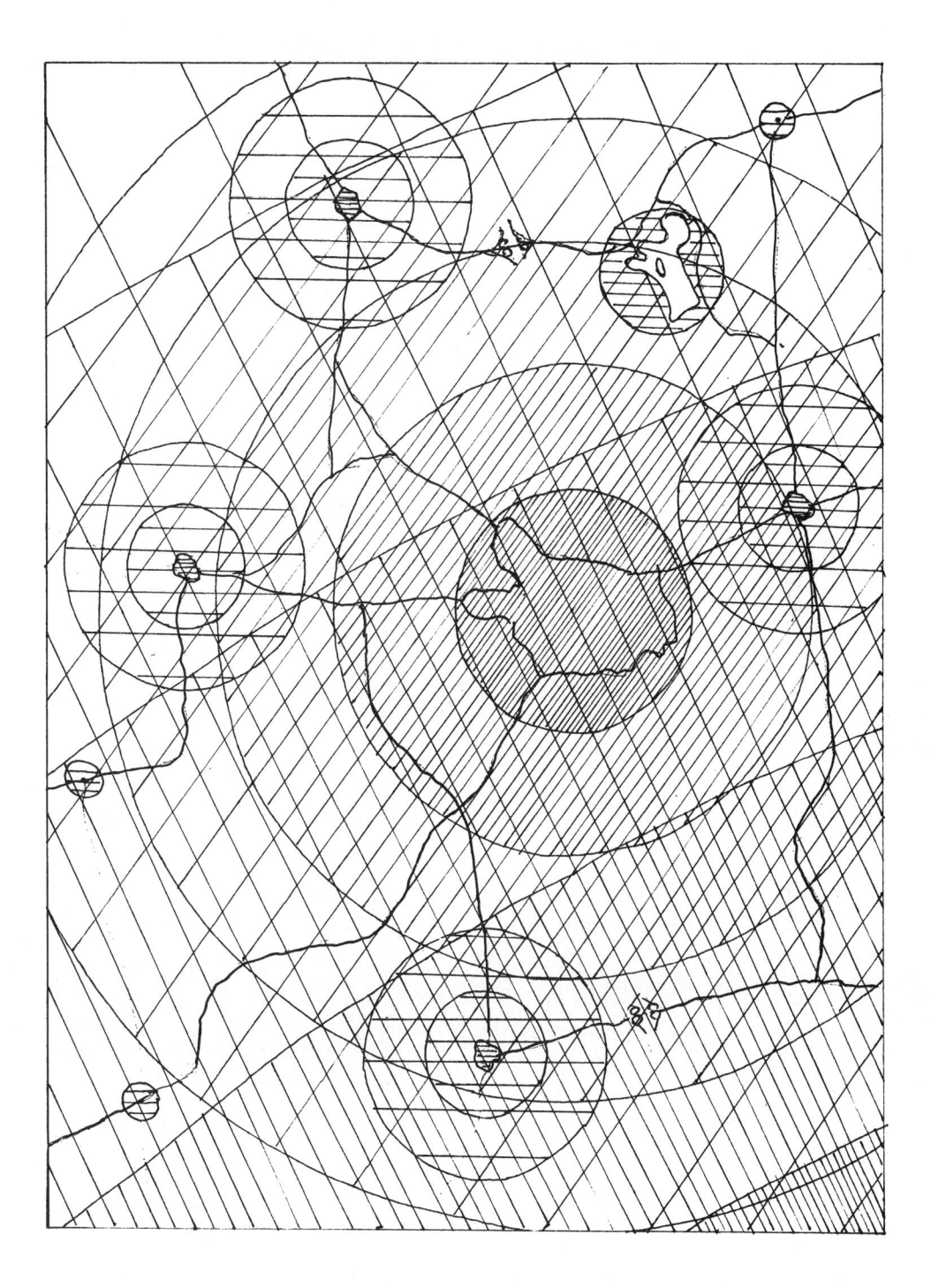

Figure 3c. Here you can see the effects of population centers on land values. The most expensive land can be found in those areas with the most cross-hatching.

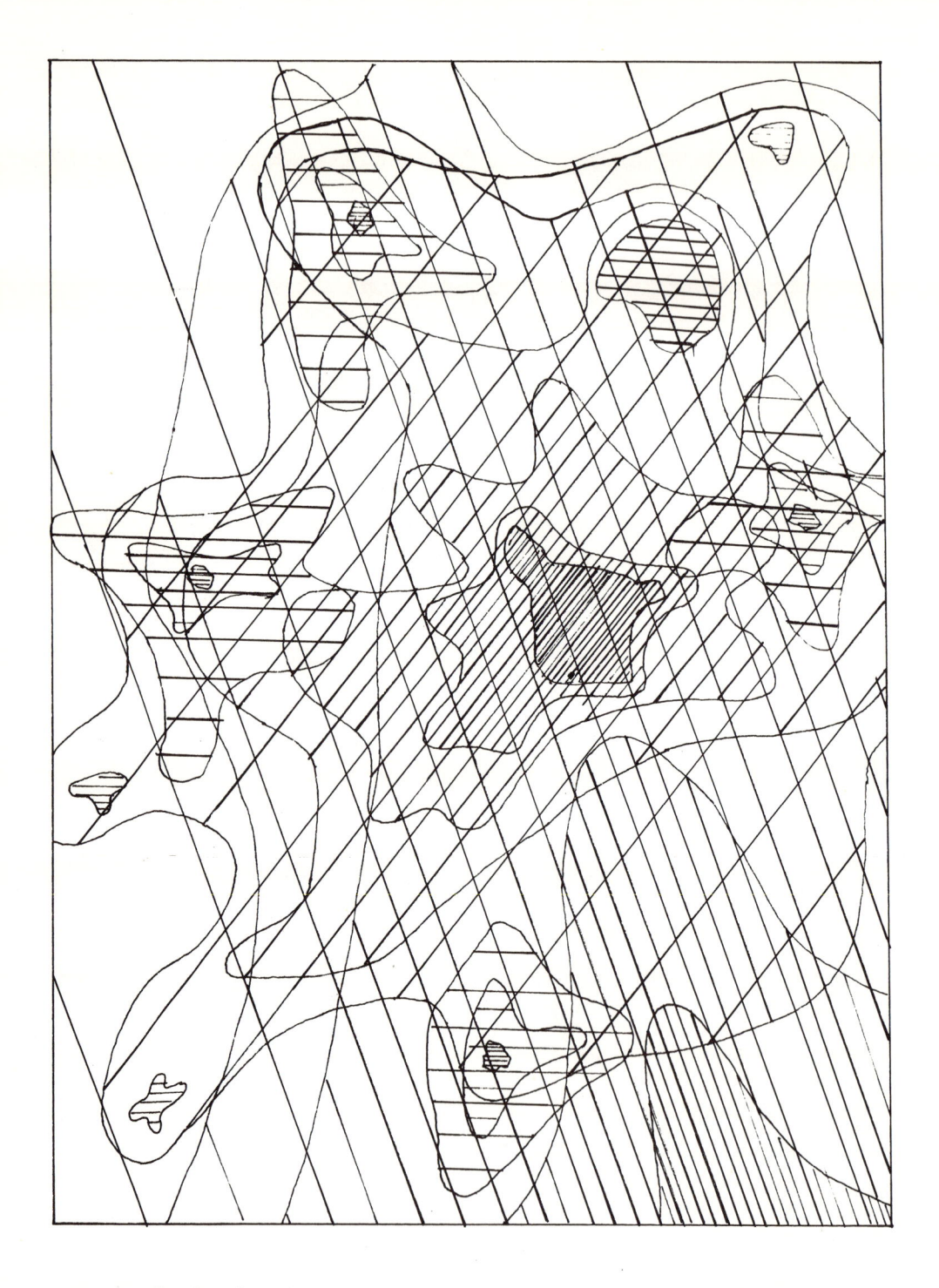

Figure 3d. This diagram is a more realistic rendering of Figure 3c, taking into account ease of access and the land's proximity to rivers and lakes.

Regulations were established calling for minimum lot and house sizes in a neighborhood, in some cases with expensive and rather wasteful minimums of two to four acres for lots and 2,000 to 3,000 square feet and more for houses. With ever more people chasing less land, and with the new reality of energy costs, such regulations are counterproductive, driving the cost of living beyond compatibility with personal freedom. Cheap energy is a thing of the past, thanks to OPEC, yet a town near where I live still has a zoning ordinance requiring that houses be built "parallel to the road." It is well known that orientation of a house—*any* house, energy efficient or *inefficient*—can mean up to a 35 percent difference in the cost of heating and cooling. The roads of the town were laid out with regard to *access,* not solar or prevailing wind orientation. A new house in this town—or an entire subdivision—may have an annual fuel cost of 35 percent more than if thoughtful siting was employed.

Building codes can be equally hard on the owner/builder's pocketbook. In 1969, the National Commission on Urban Problems estimated that "many excessive code requirements" added $1,838 to the cost of a $12,000 house of 1,000 square feet. Some of the excesses are listed below. (For 1981, these figures should be more than doubled.)

FREQUENT CODE REQUIREMENTS AND THEIR COSTS[8]

1. Foundations dug to clay when piers and grade beam would do as well $150
2. Extra number and sizing of joists 63
3. 2x4 studs supporting outside walls 16″ o.c. when 24″ o.c. entirely adequate .. 125
4. Extra sheathing 125
5. Separate siding and sheathing instead of single ⅜″ panel 330
6. Double framed 2x4s for window and door openings although single 2x4s considered sufficient 40
7. Each door and window must have own header when continuous double 2x6 atop outside wall is better 45
8. Extra door and window headers 20
9. Extra fire wall requirements in frame construction 50
10. Interior wall 4″ thick even though 2″ walls safe when non-load bearing 310
11. Subfloor must be ¾″ instead of ½″ plywood 500
12. Double 2x4 plate on all wall partitions where single member sufficient 30
13. Trusses on 16″ centers where 24″ centers sufficient 100
14. Masonry chimney when Class B flue would do a better job 150
15. Extra electric over National Electric Code when rigid conduit required 300
16. Metal conduit required for wiring when Romex (non-metallic sheathed cable) just as good 200
17. All electrical wiring to be accomplished by a licensed electrician 100
18. All plumbing, drainage, waste and vent size must be 2″ minimum 30
19. Install lead pan under all shower bases regardless of type instead of other means of water protection 50
20. Central cold air return cannot be used in heating. Each room must have its own air return to furnace 85

In *The Owner-Builder and the Code,* the authors point out that the Commission Report did not include "alternative building methods and materials which, if permitted, would provide even greater savings. Builders may save $300, for example, on the cost of a $30,000 construction if they are not required to form the foundation footing with wood. A simple trench footing performs satisfactorily and provides equivalent strength."[9]

In 1979, Malcolm Miller built a beautiful cordwood masonry home in an approved subdivision near Fredericton, New Brunswick. He reckons that he could have built the equivalent home in an area without all the regulations at a savings of $15,000 to $20,000. The required electrical work alone cost $5,000, more than the entire cost of some of the homes used as examples in this book!

Malcolm's advice now, based on his experience, is that owner/builders should avoid coded areas. I'm inclined to agree. The savings are compounded every time you turn around. The land will be cheaper in the non-coded area, the construction costs will be less and the interest cost is proportionally less if borrowing is absolutely necessary. The trade-off is that you will be responsible for the safety of your own house, which is as it should be. Use the codes as guidelines and as aids to problem solving, if necessary, but even better is the wealth of literature in libraries and bookstores which is much easier to understand because the *intent* of the recommendations is explained.

At this point I would like to respond to an argument which I often encounter and which may have occurred to you. The argument goes, "If you don't build your house to code, you will adversely affect its resale value. Banks won't finance houses which are not built to code."

Firstly, this is not strictly true. In rural areas, where codes are not so likely to be found, banks finance homes all the time.

Secondly, the owner/builder has imparted a tremendous "sweat equity" into his house, the fruit of his own considerable labor. He will derive several times his initial investment, *even if he sells his place "below market value."*

Thirdly, the whole idea of this book is to avoid the economic servitude which comes from financing. Having freed ourselves from our "gold and silver fetters," what right have we to inflict them upon others?

The Land Itself

There are many personal considerations in buying land that are peculiar to the individual, and it would be impossible to deal with all of these within the scope of this book. Obviously, if waterfront, a view, or other nonessential amenities are very important, the cost of the required land will increase dramatically—and sometimes unreasonably.

Les Scher's excellent book, *Finding and Buying Your Place in the Country* (1974, Macmillan), puts the reader in touch with almost every consideration in the land search, climate, water, soil, topography, evaluating existing structures, and so on. I shall limit my commentary to the considerations which are of particular importance to building and maintaining the dwelling unit.

Availability of Indigenous Materials

Both the dollar cost and the energy cost of transporting building materials over a long distance are unfavorable trade-offs. Suppose, for instance, a load of chipboard is carried 1,000 miles in a large truck. A considerable amount of oil is used to manufacture and transport this material. Pollution is imparted during the transportation from the oil field, during the oil refining, during the manufacture of the chipboard and, again, during the burning of the diesel fuel by the truck. The air gets dirtier. Oil becomes a little bit more scarce. Prices rise.

As we use more and more petroleum, the remainder becomes increasingly difficult to extract from the ground. The term energy cost can be understood in another way when you consider that if there is an estimated 100 million units of energy in a field, but 90 million units are required to get it to its usable state (the energy cost), the reserve is effectively 10 million units, not 100 million.

There is energy cost, in varying degrees, in any manufactured product: plywood, chipboard, finished lumber, concrete blocks, insulating materials. The advantages of using these products must be weighed against the disadvantages, in terms of dollar cost and the long-term effects of the energy cost.

I am not saying that manufactured products should not be used. They will frequently be necessary, at least for certain kinds of houses. I *am* saying to look for indigenous alternatives wherever possible. The use of rough-cut lumber from a local sawmill, for example, is cheaper in both dollar and energy cost when compared with buying finished lumber originating hundreds of miles away. Other beneficial by-products from this strategy are that the local economy benefits to a greater degree and you will generally have a better product. A rough-cut, 2-by-4 stud has a full 8 square-inches (52 cm^2) of cross-sectional area, while the dressed stud has only 5 and one-quarter square-inches of cross-sectional area.

The use of indigenous materials has other advantages outside of economics. Aesthetically, a house will be in greater harmony with its surround if it is constructed of the same materials. (Figures 4 and 5) A log house blends into a wooded area just as an adobe house seems to grow naturally from the dry, treeless plains.

We Americans are relative newcomers to this continent, and perhaps that's why so few of us look at the long-term impact of housing on the environment. In older civilizations, housing is a part of the environment—after—as well as during its life as shelter.

I have not found a book which better illustrates the use of indigenous materials than Bernard Rudofsky's *The Prodigious Builders* (Harcourt Brace Jovanovich, 1977). With regard to building permanence, he says:

These days, buildings have a shorter life span than the men who build them. The thought that a house might serve a family for several generations, and serve it well, has no currency. We accept premature architectural decrepitude as a matter of course. The culminating point in a building's life is reached the moment it is finished and its photographs taken for publication. In a way, this parallels the once popular belief that a person's days of youthful exuberance and venturesomeness end with marriage.[10]

Rudofsky points out that even animals—the term "owner/builder" would seem redundant to a honeybee—do better than people in this respect:

Beavers' consolidated bulk of work will stand up to the ravages of time for thousands of years, a record rarely matched by man-made constructions, perhaps because beavers are eager to provide constant supervision and labor for repairs, while man is not.[11]

A few pages later in the text, Rudofsky calls attention to a large termite mound in Rhodesia which, "having been submitted to the most exacting tests, owned up to its age of 700 years. A pickax could not dent it; it had to be blown up with dynamite to make way for a road."[12]

Rudofsky's explanation for the success of animal "architecture" hits close to home:

Granted, animal builders work under enviable conditions. Unhampered by red tape, and innocent of profit motives, with an incalculable backlog of practice at their disposal, they often attain perfection by simply following their instinct.[13]

Again, building codes and profit motives—often one and the same thanks to successful lobbying—surface to complicate the relatively simple matter of shelter. But natural building skill and lack of building codes by themselves are not sufficient to explain the quality of ancient dwellings, their beauty and their durability. The use of local materials is important. When you, like the animal, simply reshape materials at hand primarily because they have the ability to survive the local conditions of sun, wind, rain and frost, well, you stand a better chance than if you were to bring in materials—even natural materials—from somewhere else. An adobe house makes as little sense in Alaska as an igloo in New Mexico. Neither is likely to last very long.

A good example of the use of indigenous materials is the construction of the "soddies" of the American North-Central Plains during the last century. Lacking any other building materials, these settlers built comfortable houses of the very ground itself. Years later, when transportation made cheap wood available, cheap enough that barns, fences and outbuildings could be made of it, many perceptive builders continued to use sod *because it made more sense in terms of comfort and durability.*

Sod house historian Roger L. Welsch says, "The frame house was painfully vulnerable to prairie fires, severe weather and wind. Wood burns, shrinks, swells, rots and can be eaten by insects and rodents. Sod resists all of these."[14]

Figure 4. An exterior view of Log End Cottage.

Figure 5. And an interior view of Log End Cottage.

With regard to the selection of land, consider the kind of house you want and the suitability of the available materials to the climate. If the most practical of the indigenous materials in your area is wood, but all of the woodlot was clear-cut a year before the land came up for sale (a common occurrence), it should be obvious to you that its value is greatly diminished if what you want to build is a log cabin. On the other hand, the logging slash left behind might be suitable for a cordwood masonry house. This could prove a bargain all around if the land value was so greatly diminished during the clear-cutting.

INDIGENOUS ENERGY SOURCES

Shelter is one of the "big three" on Thoreau's list of "the necessaries of life," fuel and food being the other two. (Clothing, the fourth means of retaining body heat, can be as expensive—or as inexpensive—as individual taste dictates.)

Fuel, including the potential for electrical energy, can be a valuable bonus on your land. This potential may or may not be reflected in the price of the parcel, depending on the degree of perception of the seller. The obvious fuel source for burning is wood. In the case of clear-cut land, the logging slash left behind might be of considerable value as firewood, but the question is whether or not it will last until the new growth is ready for harvest. This will depend on the degree of "clear" cutting, and the energy-efficiency of the home to be built. Coal, oil and gas, however, are not likely to be economically harvestable by you yourself.

A future owner/builder visited my wife and me recently and told of how a gas company had driven a well on his father's land where our visitor planned to build his own house. At first, the deal seemed excellent: one-eighth of the gas from the well plus a cash payment. But six acres would be lost forever by the need for a salt-water covering for the land near the well. Not so good.

A cheap source of coal nearby, even though it is not directly on the land being considered, might be a positive factor. However, like the gas well, it is of a limited benefit and the obvious negative factors of living in a coal or oil field should be considered.

A well-managed hardwood lot can yield a half cord of wood per acre per year, which means that six acres can perpetually supply a house which requires three cords of wood per year for heat—and the woodlot will actually improve in quality! For fuel, hardwood has a higher value than softwood, whereas softwood is preferable for building. Many people in the West heat with softwood, however, due to hardwood's unavailability, and many a fine home has been built of oak and even poplar.

Water power potential will usually be reflected in the cost of the land, even if by way of other criteria, such as a river's value for fishing and recreation. But water power systems are likely to be quite expensive, unless, like a friend of mine in Massachusetts, you find an old mill at a good price and have several years experience in running a foundry so that you can cast your own turbine blades.

Wind is a potential energy source which rarely influences the selling price of land, probably because people who use windpower are still so few and far between. Good wind sites, however, are not as common as you may think.

I was cofounder of a company in northern New York state that sells and installs wind machines. I would often hear people say, "I've got a great wind site; there's always a breeze." They didn't realize that the potential power in the wind increases as the cube of the wind speed. For example, a 20-mile-per-hour wind has 8 times the potential of a 10-mile-per-hour breeze. The bulk of the power made by any wind machine occurs during the time of the "power winds" and the "steady breeze" experienced at certain sites—the edge of a small lake, for example—may not help very much in supplying power. One good frontal storm, on the other hand, might be sufficient to charge a small battery system.

If wind power is an important consideration, make sure that the site will meet your needs. At the very least, get a visual analysis from someone who is familiar with wind power. If he or she says, "Forget it," forget it. I know of no dealer in wind machinery who will tell you you've got a good site if you don't, just to sell a wind machine. In a borderline case, you might be advised to have a three-month analysis done with a recording anemometer to find out the average wind speed at your site. The figures derived from such a short test should be compared with the figures for the same period recorded by the nearest meteorological station or airport. Then the two sets of figures should be compared to the average for those three months over the last 20 years. This double comparison method should make up for the gaps in an abnormal testing period.

The most economic approach is actually to buy the testing equipment and set it up yourself. A wind machine dealer will charge you the approximate worth of the anemometer for a three-month site analysis, although reputable dealers should deduct the cost of the analysis if you buy a wind machine.

When you are finished with the anemometer, you can sell it at a small loss through a classified ad in your local paper or through *Wind Power Digest,* 54468 CR31, Bristol, IN 46507. The *Digest,* to digress, is also a good place to start looking for used wind machines, which are very often—but not always—a good buy for the low-budget wind energy aficionado. (I do not advise building your own wind machine from scratch unless you are the kind of person who would be quite happy also building your own car from scratch!)

There are two different kinds of systems for making and using electricity from the wind. The best system, economically and philosophically, is the battery storage application, also called the "remote site" system, although it is not limited to remote sites. With this system, the user is responsible for budgeting his or her own supply of electricity as stored in the batteries.

At the Roy homestead, we store 720 ampere-hours of power at 12 volts, which we can stretch to about three weeks' use without wind, if need be. But, understand that I am talking about totally different consumption patterns from those of the "average" household.

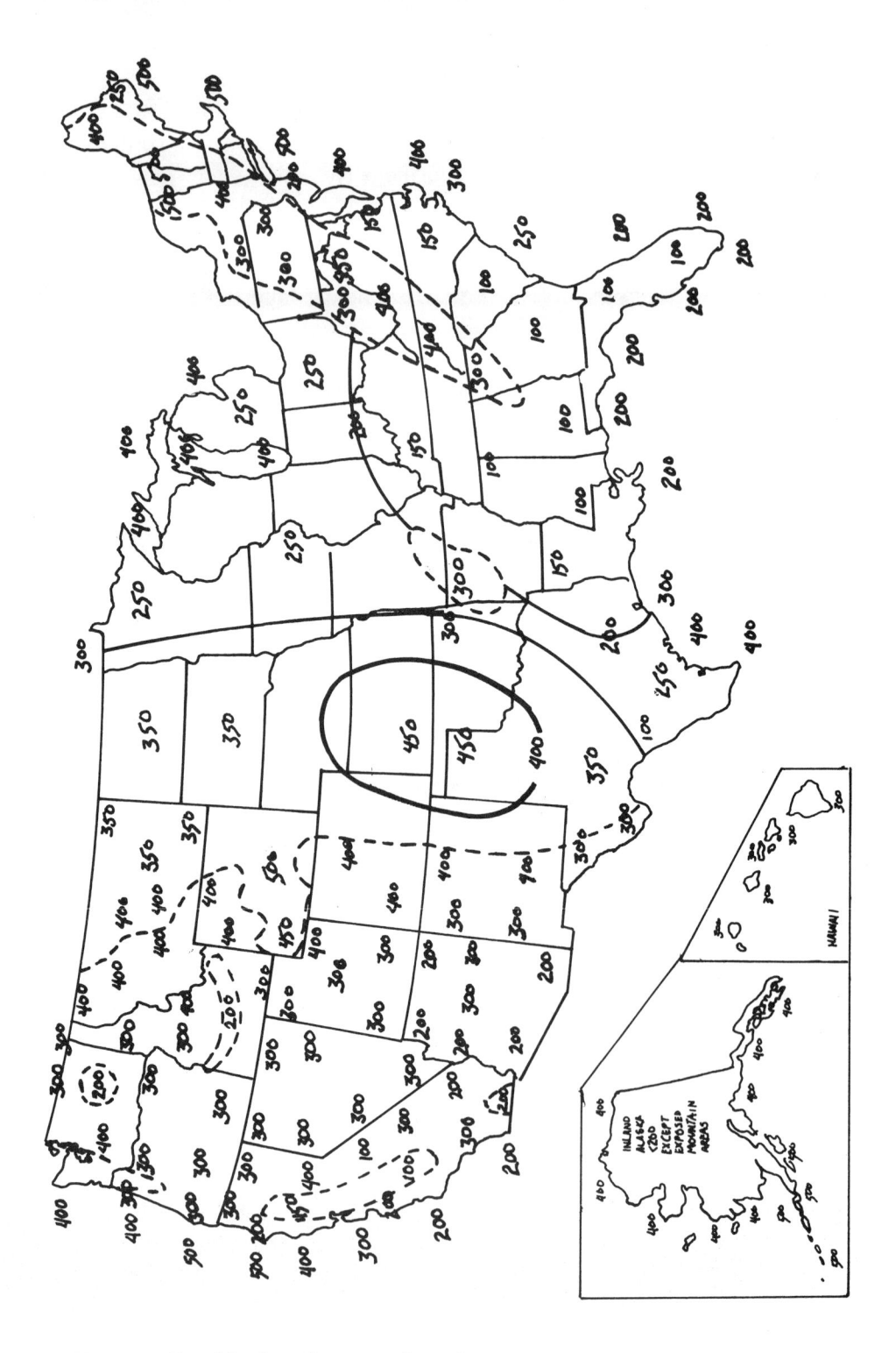

Figure 6. Use this chart for annual wind power averages across the country.

We use electricity for lights, stereo, television (a 12-inch, black and white model) and a small water pump. Twelve-volt d.c. lighting is between two and two-and-a-half times more efficient in terms of candle-power per watt than 110-volt a.c. power. We cook with wood and gas, and run a small refrigerator on gas, although we could run it on 12-volt power with a larger wind system. We use a 500-watt Sencenbaugh windplant on a 60-foot guyed tower. Other, larger, wind machines with proportionally larger storage systems will accommodate greater consumption and more kinds of appliances, although appliances with heating coils, such as stoves, water heaters and space heaters, are not advised because of their large electrical requirements. Inverters allow you to use virtually any appliance, but there is a certain amount of power wastage while the inverter is in use.

The other kind of wind system is the "utility interface" system, also known as the "on line" or "reverse feed" system. With this system, the wind machine makes electricity that is synchronous with that made by the power company. Some machines do this right in the generator, while others require a "synchronous inverter," a very expensive piece of equipment, to do the same thing.

When power is made, it can be used directly in the home without affecting your monthly power bill in any way. When the windplant is making power in excess of the household draw, an electric meter records how much power is being put into the power grid. When the wind is not blowing, which is most of the time at most sites, you must pay for your power at the regular rate. Unfortunately, the consumer is credited at wholesale rates for the power he supplies and charged regular retail rates for the power he consumes. And a rental is charged for the second meter. *And* the consumer's windplant cannot be used—in fact, the blades will not turn—during a power outage! (This safety feature prevents a line repairman from being zapped with a surge of someone's homemade power during an outage.) With regard to the price of land, remote sites (those not near power lines) will be less expensive.

For these economic reasons, and for the extra freedom that comes from not having to deal with a regular power bill, I prefer the remote site system. Of course, the remote site system places a greater responsibility on the householder to conserve, but this is as it should be. As long as we're going to all this trouble to save on shelter, why not tackle the next biggest living expense, fuel, at the same time?

SOLAR POWER

The potential gain from solar power is largely a function of geographical region. In almost every part of the country, solar hot water heating is economical, even if this only meets a portion of the fuel cost. Obviously, the more favorable the solar index (Figure 7), the shorter the payback period for a hot water heating system.

Photovoltaic cells for electrical power continue to come down in price, although they are still a bit on the expensive side when compared with wind power. If I were building in the Southwest, or if wind power was not possible

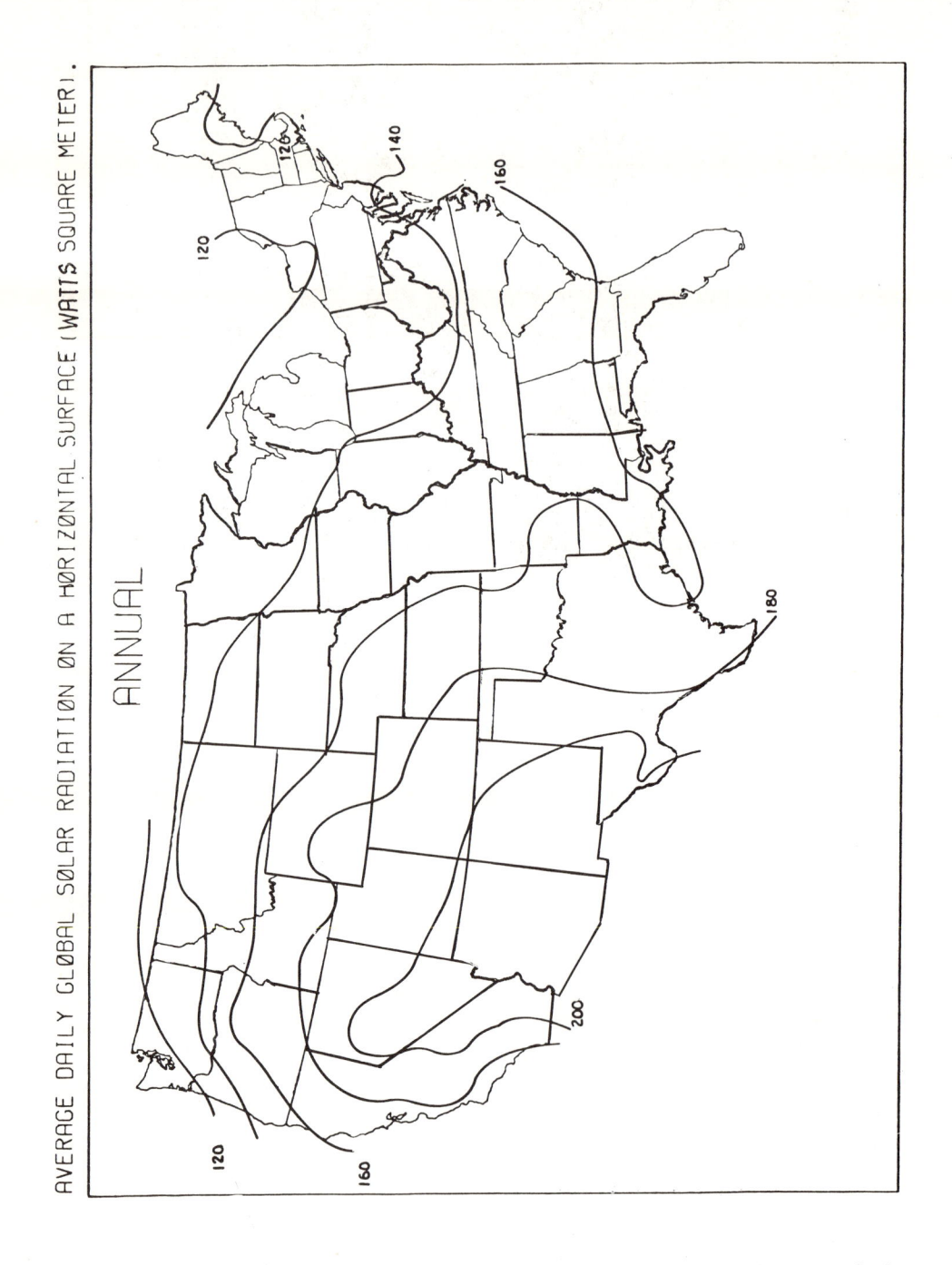

Figure 7. Solar hot water heating is almost always economical, particularly where the solar indexes are most favorable. Use this chart to find them.

due to a poor wind site, I would investigate photovoltaic cells as a potential source of electricity *now* (1981). My personal belief, based on my experience in the industry, is that mass-produced photocells will make electricity at a lower cost than small wind systems by 1983 or 1984. The maintenance cost is already lower. Applications of homemade electricity with photocells are very much the same as those with windpower, already discussed.

The potential for savings from alternative sources of energy is not limited to energy. Land prices will be less wherever the cost of bringing in power is prohibitive. And savings on your land price might be enough to pay for a good wind or photovoltaic system. (Don't forget to find out if the 40 percent energy credit on federal income taxes is still available—it still was as of 1981—as well as any state credits.)

Food Production

VEGETABLES

Food production is another case for potential on-going savings. My personal experience might be of benefit to you, particularly if you don't have previous gardening experience. One of the reasons that my wife and I left Scotland in 1975 and moved to northern New York was that we had only one-quarter acre of land, less than half of which was fit for growing vegetables. We didn't know then that one-tenth of an acre is more than enough to grow all the vegetables required by a family of four, if the raised bed, or "bio-dynamic French intensive" method of gardening, is employed. Using this method, which will be explained below, we now grow all the vegetables we can use—and then some—in a space about the same as what was available to us in Scotland. Sometimes, people looking for land insist that they must have "some cleared land—an acre at least—for a garden." We said the same thing in 1975. Now we realize that even wooded land can be made fine for vegetable gardening at little expense.

French intensive gardening involves the use of several "raised beds" of about 4 feet in width—the maximum width allowing the gardener to reach the center from either side—and virtually any length, although 8 feet is common. I built one raised bed 32 feet long with occasional "crosswalks" to facilitate moving around the garden. The depth of the good soil that is turned each year, should be at least 12 inches, although this varies somewhat with the crop. Beds can be either permanently framed with stone, blocks, or wood, or can be shaped with a rake and a spade. We like the "permanent" beds, but have seen many excellent gardens of mounds reshaped each year at the time of turning. One of the prettiest gardens—and most productive on a square-foot basis—that I have seen was virtually carved out of the edge of a recently harvested woodlot. Using stones, logs and the natural contours, the gardener built dozens of small, easy-to-manage beds, connected by a winding path. Land which had been quite badly scarred became beautiful and productive.

The advantages of raised beds are:

1. Less land is needed. Intensive planting is possible as no spaces between rows are required, yielding much more produce per square yard of garden.

2. Seeds or seedlings are spaced so that the young plants form a living "green mulch" over the bed, discouraging weeds and helping to retard moisture loss through evaporation.

3. Gardening is three-dimensional: leafy vegetables are alternated with root vegetables so that both the surface and the subsurface is productive.

4. Each bed can be tuned to the proper pH factor (acidity vs. alkalinity) for the particular vegetables to be grown in that bed. Lettuce likes sweet soil, for example, while strawberries prefer acidic soil. Certain special mineral requirements also can be economically satisfied by concentrating them where needed.

5. Much less watering is required because of the lack of runoff and reduction of evaporation wastage. This advantage is greater with the permanently bordered raised beds, as opposed to the mounded method.

6. Much more economic use is made of mulch and compost with intensive gardening. Good soil is built up faster.

7. No rotary tiller or cultivator is required. The beds are easily maintained with a spade and small hand tools. The raised beds are easier to work on because they are 8 to 10 inches above the permanent walkways.

8. Easier bug control.

9. Aesthetically superior. Neater and tidier with the permanent walkways. Very little weeding required.

10. In combination with a built-on protective cover and the use of rigid foam insulation around the inside edge of the bed, growing seasons can be greatly extended. Leandre and Gretchen Poisson, who pioneered this technique, grow certain vegetables year around this way in snowy New Hampshire.[15] (See Figure 8 for my own sketch of this process.)

The best reference work that I know of on this subject is *How to Grow More Vegetables* by John Jeavons (Ten-speed Press, Berkeley, 1979).

LIVESTOCK

If sheep, goats or cows are a part of your food production plan, the land will have to provide pasture. In the case of cows, additional land adequate for the growing of winter hay is needed. Growing winter feed for goats and sheep is a dubious economy for the homesteader, unless a fairly large flock is kept. Buying winter feed from a local farmer is probably the better strategy.

Converting a genuine woodlot to hayfield is not recommended as a sound economy, either. What is viable is to reclaim formerly cleared lands. Just make sure encroaching treestalks do not exceed an inch or so in diameter. (Three inches if a tractor is available, and it will have to be available, if growing winter hay and grain is a part of the plan.)

In any case, the land must be evaluated for such considerations as depth and quality of the soil, frequency and size of stones and drainage. In northern

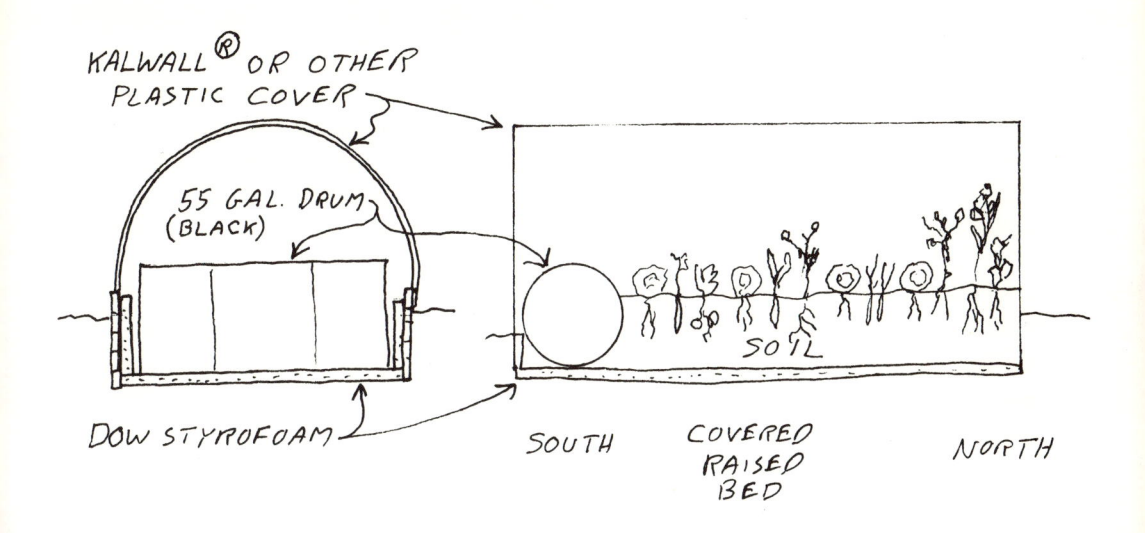

Figure 8. A raised bed combined with a protective cover and rigid foam insulation can greatly extend your growing season, among other advantages.

New York, a good meadow will support about six sheep or goats per acre as pasture. Cows will require from one to five acres *each* as pasture, depending on the quality of the land. In addition, beef cattle will require an additional acre of good hayfield to produce winter feed; two acres each for dairy cows. Obviously, these figures are different in areas with a longer growing season. If you are an inexperienced owner/builder but find the farmsteading life-style appealing, have a long talk with the local county agricultural agent or cooperative extension agent prior to buying your land.

Pigs and rabbits do not require very much land at all and are considered by many modern homesteaders to be the best livestock for economical meat production. The manure from these animals is particularly effective in the garden. Chickens supply meat, eggs and strong manure and can be raised on a wooded lot.

We eat little meat and do not keep livestock at the Roy homestead, although we are not vegetarians. For us, the curtailment of our freedom to travel is not a good trade-off for the admitted economic benefit of keeping small animals. If personal and economic situations change, we would not be ill-disposed to keeping rabbits or hens.

OTHER FOOD PRODUCTION

Maple syrup production is particularly viable in our area, and thrives as a cash crop in northern New York and New England, Wisconsin, Michigan and the Appalachian Mountain region. A mature maple grove—called a "bush"—adds considerably to the value and price of land. It can also be a source of income, either by producing the syrup and sugar yourself or by letting out the bush to a commercial boiler.

If there are young sugar maples in your mixed hardwood forest, they may not have increased the land price, yet, by proper woodlot husbandry, they may produce a viable sugar bush in a very few years, a good investment of time and money.

On the other side of the coin, sugaring equipment is expensive, unless you are fortunate in finding a bargain situation in which someone either is closing down a sugar house or moving up to a bigger boiler. And sugaring is lots of hard work. You may think that $20 per gallon for maple syrup is outrageously expensive, but I can assure you from my own experience that sugaring is not a get-rich-quick scheme. We do it for the syrup, which we use at home and give as Christmas gifts, and, yes, we do it partially for the romance of it. After a long and tedious North Country winter—especially in March when "cabin fever" reaches epidemic proportions in our area—the first sap run is a joyous event. That otherwise dismal month of cold rains, mud and wind is cheered by the cozy atmosphere of the sugar house with its blazing wood fire and sweet smell of boiling sap.

Beekeeping is probably a more economic proposition for the production of sugar than mapling, and suited to a wider variety of landforms and climates. We do not keep bees, for no particularly good reason except, perhaps, that we do have bears in our neighborhood. Still, two of our neighbors keep hives and have not been bothered by our friends in the woods.

A producing orchard—apples, pears, peaches, whatever—will add tremendously to the land price, but a few abandoned "scrub" apple trees or the like will not, especially when the fruit looks small, scabbed and wormy. Hopefully, what the seller doesn't realize is that these old trees and their offspring can be very valuable to a homesteader. They have strong, long-established root stocks and it isn't difficult to learn how to graft scions of McIntosh, Golden Delicious or any other desired apple onto these established trees. This is a fast way to establish good fruit production and has the advantage of using an established, hardy root stock. (See Country Wisdom Bulletin A-35, *Grafting Manual,* Garden Way Publishing Co., Charlotte, VT 05445.)

Marginal Land

When land which has been raped by people of its resources—wood, topsoil, gravel and so on—comes up for sale you might not consider owning such a parcel because it looks a mess and seems so barren of possibilities. An attitude like this is often reflected in a very low asking price.

"Right," you say, "and I don't want that piece of land either." But don't dismiss this land too lightly, especially if your primary goal is to find a cheap building lot, and not to produce the various food, fuel and indigenous materials already discussed. So often we see heavy machinery turn a pretty lot into a wasteland before construction of the house anyway. This is an example of the prevailing policy in America of fitting the site to the house, instead of the more thoughtful and gentler attitude of fitting the house to the site. Bulldoze, build, landscape. Trees are destroyed; topsoil is homogenized with subsoil; additional expense is imparted to already expensive land to get it to the point where "convenient" building can take place. The use of already decimated land may enable you to afford property closer to town and at a lower price than that of "prime" land. The cost of landscaping usually can be subtracted from the savings of the land cost, especially if the work is done by you yourself.

There is another positive aspect to buying marginal land. For so many years, American corporations and individuals have taken, taken, taken from our land, and given so little back in return. Instead of pillaging more of our land for convenience and short-term gain, let's reverse the trend wherever we can. A lot to build a house on need not start out as a part of Eden; we can accomplish the transformation ourselves. A house requires, first and foremost, space to build it on, not scenic views and streams and prime agricultural land. Fellow advocate of underground construction Malcolm Wells tells about the building site of his first underground office:

> Cherry Hill, New Jersey, is a lavishly rich, trashy suburb of nearby Philadelphia. My office is a tiny place on a tiny lot, wedged between a freeway and a sewer. When I bought the property (for $700!) all I could see were a few scabs of old asphalt on a patch of barren subsoil. It was dead; all the way down the scale from forest to woods to farm to suburb to abandoned highway construction yard. Now, five years later, it's almost a jungle, even though no topsoil and no fertilizer have ever been used, and a building now underlies almost half the root space. The secrets: plenty of mulch and a few key starter-plants . . . Now, when we tell our clients how to find choice building sites, we always urge them to pick the *worst* ones, not the best, as we were always taught to do. Now people can see for themselves how easy and how gratifying it is to restore a bit of this trampled continent.[16]

My wife and I are building a new earth-sheltered house on the same hill where, in 1975, we joined with some friends (several of whom recount their sagas in this book) to start a community of owner/builders. It is a beautiful hill, with woods, meadows, springs and streams. If an eyesore exists in the neighborhood, it is a 2-acre piece of land from which gravel was removed many years ago to build the dirt road which leads to the hilltop from the state highway below. The excavation of gravel ceased when sand was struck, at a depth of about 4 feet, so the extraction proceeded laterally, turning a large area to waste, where almost nothing grows and the sand is whipped up

Figure 9. Believe it or not, the ideal building site is dying land that is just waiting to be restored. The prospects: silence, privacy, wildlife, clean air and water, more topsoil, less erosion and lower fuel bills. This drawing is reproduced here with kind permission of Malcolm Wells of *CoEvolution Quarterly*.

by the wind. It is not what you would think of as the ideal building site. But, as you know, the repair of the one weak link in a chain has a more positive effect than the strengthening all of the other links combined. So, too, the restoration of this one piece of devastated land will mean more to the visual and ecological amenity of the hill than any amount of landscaping where Nature has already set the standard. This project will be discussed at greater length in Chapter 5.

Buying Land Cooperatively

In 1974, Jaki and I were living in the Scottish Highlands. As Jaki was British and I was an American with permanent resident status in Britain, we had the option of staying in our cottage with its quarter-acre of land, or living in the United States. Land pressures in Britain combined with a very restrictive nationwide building code (not to mention strict local codes), making pursuit of the self-reliant life extremely difficult in Britain.*

We travelled to the United States and bought a used camper van. During the next four months, we traveled 15,000 miles and visited 35 states, searching for land and enjoying a belated honeymoon. We found "the right piece" in New York, 15 miles from the Canadian border. The story is worth relating, as it illustrates an important point: not all land for sale is advertised.

One day, while travelling a country road near Plattsburgh, New York, we commented on the beauty of the gently rolling meadows with hardwood forests at their far side. We noticed that most of the land was posted against hunting by a single landowner. Soon, we came to a mailbox identified with the same name. We stopped and asked the man, a semi-retired farmer, if he had any land for sale. He said, no, he was saving his land for his family, but knew of some abandoned farm land "up on the hill" that was for sale by someone else. He even took us to see the land, a mile off the main road. Pleased with our first impression, we phoned the owner, arranged a meeting to look at the land properly, and ended up buying it. The land was not marked "FOR SALE," nor did the owner advertise in the papers. You had to hear about it through word of mouth. The lesson: in an area which appeals to you, stop and ask the local people if they know of any land for sale. Many country farmers will not put a "FOR SALE" sign up, lest their neighbors think they are being forced to sell because of some failure in their farming enterprise.

CoEvolution Quarterly editor Stewart Brand disagrees with my advice of

---

* AUTHOR'S NOTE: If you are a member of a town or a county planning board, you may want to shout at me, "You fool! Can't you see that it was the land pressures themselves, the overcrowded conditions of the British Isles, that made restrictive planning necessary to protect the remaining countryside?"

In southern England and Wales, this argument has a certain validity. The available resources-to-population ratio has been allowed to deteriorate to the extent where Draconian measures seem justified. The Scottish Highlands, on the other hand, are sparsely populated and the land is controlled by a very few large landowners (often absentee landlords), few of whom are Scots. The local people are crowded into the few towns, while the vast majority of the land is devoted to such pursuits as grouse hunting and salmon fishing. Whatever little benefit deriving from the "husbandry" of land in this fashion rarely percolates down to the people who actually live in the Highlands. Yet, it would be as wrong to abolish all planning as to turn the whole planet over to it. There must be flexibility and intelligent examination of each situation, but all too often commissions formed to determine the best use of land fail to consider the option which is sometimes the appropriate one: "No planning is the best planning." At the very least, planning (and *building*) codes should be *locally*—not *nationally*—applied.

the previous paragraph. He says, "Wandering around in pretty country, falling in love with various sites, is no way to get land. It always either belongs to the government, who won't sell, or to some guy who won't sell."[17] I suspect that the local pressures on land already discussed have a lot to do with the success or failure of knocking on doors.

Jaki and I had no desire to live a hermit's existence on an isolated piece of land. The seller had another, much larger, parcel across the road from the place which we bought, and he gave us a one-year written option to buy the additional land. We returned to Scotland and placed the following message in a national magazine read by people interested in self-reliant living:

> We're a young couple moving this spring to a secluded 244 acres near Plattsburgh, New York, to build a house and grow some food. We're looking for several peaceful folks—to come in with us on an individual ownership basis—who have respect for nature and are willing to help each other out as necessary. We own 64 acres and have an option on the rest, which could be bought at $125 to $175 per acre. Land rolls gently, has mature wood, good meadows, clean streams and no buildings. Access by town road, plowed in winter. Details by return of post. Tell us about yourselves.

During the next few months, we received over 90 replies, mostly from people looking for land, but also a few from people in the Plattsburgh area who just wanted to say "hello" and invite us to stop in to visit when we got to America. This was an unexpected bonus; some of these people have become good friends.

We replied to each enquiry, and exchanged several letters with many of the correspondents. When the list was pared down to a dozen seriously interested families, we all agreed to meet on the land in April of 1975.

Ten families ended up joining with us on the hill as a result of our endeavor, and other families have since joined the community on riparian land. We now have over 40 people.

Each family obtained a warranty deed for their parcel, and, in most cases, five-year mortgages were arranged through the owner at a low rate of interest. Typical land payments were $500 to $1,000 per year, depending on acreage and ease of access.

By dividing up a large parcel of land, at a relatively low cost per acre, we all benefited over what individual 20-acre parcels would have cost.

Before a group jumps in and buys such a property, you should make sure that you all understand the subdivision restrictions—if any—that apply in your chosen town or county. Where we settled, for example, a property divided into more than four pieces came under the authority of rather stiff subdivision restrictions. Luckily, the land which the group bought was in three parcels and we were able to divide them into two lots, three lots, and four lots respectively.

Each family on the hill owns its own piece of land. There are other options involving co-ownership, but these option agreements must be entered into very carefully indeed. I would not advise a group of haphazard strangers—such as ours was at first—to attempt such an arrangement. We did visit a community in Arkansas where a group of friends joined together in the buying of a large parcel of land. These people had been friends all through high school and college and knew each other well. Their co-ownership was a success. Also, communities with a common religious bond stand a good chance of avoiding individual conflict. Some co-ownership possibilities are:

TENANCY-IN-COMMON

Each buyer shares an undivided interest in the land. In the case of the death of one of the co-tenants, his share is passed on to his heirs.

JOINT TENANCY

Similar to tenancy-in-common, but the co-owners have the "right of survivorship," which means that the interest of a deceased member of the joint tenancy passes to the other owners.

TENANTS BY THE ENTIRETIES

Applies only to a husband and wife, and only in certain states. Each marriage partner automatically has the right of survivorship. Jaki and I are "tenants by the entireties."

CORPORATE OWNERSHIP

A group can form a corporation solely for the purpose of owning land. Agreements can be made among the owners of the corporation. The corporation has all the legal aspects of a single person. This strategy should be entered into very carefully and I would only consider it if subdivision were not possible. Any business dealings should be done precisely, clearly and legally.

THE UNINCORPORATED NONPROFIT ASSOCIATION

This is not a rock group! One of the members is appointed the trustee of the association in order to take title, but the association retains liability.

CASUAL CO-OWNERSHIP

Emphatically not recommended, especially among relatives or casual friends.

For more information on these co-ownership possibilities, consult a lawyer in the state where the land is to be bought and read attorney Les Scher's *Finding and Buying Your Place in the Country* (Macmillan, 1974), Chapter 32. Mr. Scher wisely suggests that an owners' agreement be drawn up and signed by the several parties. An excellent "Model Owners' Agreement" appears on pages 351-353 of Mr. Scher's book. I know of a successful co-ownership community in Vermont that used Mr. Scher's model, with a few deletions and additions of their own.

KINDS OF DEEDS

Basically, there are two kinds of deeds. One, the Full Covenant and Warranty Deed (also called, simply, a Warranty Deed) gives the buyer the assurances that the seller in fact owns the land he is selling. He warrants it so. A Quitclaim Deed, on the other hand, transfers only the rights which the seller has in the land, which might be none at all. It is quite legal to offer for sale a Quitclaim Deed on the Brooklyn Bridge. Les Scher strongly admonishes land buyers to reject Quitclaim Deeds out of hand.[18]

TERMS OF PURCHASE

Unless the land is bought outright—which is the best method, if you can afford it—the bargain usually will involve a mortgage, deed of trust or a land contract. By the mortgage or deed of trust methods of buying land, the buyer receives a deed to the property, which he or she records. In the case of the deed of trust, a trustee will hold the deed until the land is paid for. By the land contract method, no deed is delivered to the buyer until all payments are made. This is risky business and is not advised. The terms may seem favorable, but the risks are high. If a land contract is absolutely the only way you can afford to buy land, read pages 219-221 in *Finding and Buying Your Place in the Country* before signing the contract. Although he does not recommend the purchase of land "on contract," Les Scher lists several essential protections that should be written into the land contract.

A mortgage is a legal agreement of financing. The owner of the land may "finance the deal" himself and hold the mortgage, or a lending institution may hold the mortgage. My experience is that financing by the owner is very often less costly in terms of interest than a bank mortgage. Many sellers are willing to "give terms" because they benefit on their income tax by not taking all the money at once. Holding the mortgage themselves is better for them, and may be better for you.

As for saving money for the land, read the next chapter, entitled "The Grubstake," which is a detailed discussion of this very subject.

3

The Grubstake

The word "grubstake" originally referred to money or provisions advanced to a prospector in return for a share of his findings. "Grub" was advanced for a "stake" in the claim. Nowadays, the term commonly refers to monies laid aside for the purchase of real estate. Basically, there are two strategies which can be pursued with regard to owning your own home. I call them the "full grubstake" and the "land grubstake."

The Full Grubstake

This strategy requires that you lay by—or already have amassed—all the money needed to buy a piece of land *and* build a house upon it. This is nice if you can do it. Its primary advantage is that you can "strike while the iron is hot." Enthusiasm carries over from the purchase of the land to the construction of the house as there is no saving/waiting period. In addition, a complete break can be made from your current mode of existence, a boon when your current mode isn't really what you want out of life anyway.

If the land is close by to your place of employment, money for land purchase and building materials is sufficient. If a move is anticipated and outside employment cannot be counted upon right away, additional money for ordinary living expenses during construction also will be necessary. Obviously, you'd better consider where your living is to come from *after* construction of the house. (It's amazing how many owner/builders omit consideration of this logical factor.) Even though the cost of living reduces greatly when shelter costs are eliminated—and even more if food and fuel expenses are attended to simultaneously—it is still necessary to come up with a certain amount of

43

money to live. The whole exercise has been a waste of time if you cannot earn the "paper costs" of living, such as mandatory insurance, land and school taxes, registrations and licenses, which seem unavoidable. And, too, a certain number of luxuries seem almost to have become necessities in today's society: recreation, entertainment, labor-saving devices.

You already may have built up enough equity in your currently mortgaged home to sell out and clear enough cash to build your own home, saving the economic servitude of the remaining years on the mortgage. Or, you may already own your own home free and clear, but want to make use of the economic benefits of owner/building in another way—for retirement, as one example.

A typical scenario: a couple in their late forties live in a large, fashionable, but energy-inefficient home which they own or almost own. Appreciation on their original investment has been good. Family income is steady. But the house is too big for them now that the children have left. And the new reality of energy costs looms heavy on the landscape. They're interested in energy-efficient homes, such as passive solar and earth shelters. Still young enough to tackle a building project, they see it as a personal challenge and adventure. So, they sell their home for $80,000, buy a piece of land and build a sensible and personally-attuned home at a combined cost of $40,000, putting the extra toward retirement. Nice.

If you are starting from "scratch," the problems connected with laying up the full grubstake may outweigh the benefits. Inflation is an enemy. Land and building material prices often rise even faster than the general inflation rate, which most economists see running unabated at 10 percent and perhaps more, throughout the 1980s. This has to be so, it seems to me, as long as a government continues to spend more than it takes in, relying on the printing presses to make up the difference.

Consider the following example.

Year	Amount Saved Laid by	Interest	Total	Amount Needed (10% Inflation)	Need to Save
0	0	0	4	10,000	10,000
1	2,000	60	2,060	11,000	8,940
2	2,000	183	4,183	12,100	7,917
3	2,000	311	6,494	13,310	6,816
4	2,000	450	8,944	14,641	5,697
5	2,000	597	11,541	16,105	4,564
6	2,000	752	14,293	17,716	3,423
7	2,000	918	17,211	19,488	2,277
8	2,000	1,093	20,303	21,437	1,133
9	2,000	1,278	23,582	23,581	−1

A couple calculates that they need $4,000 for land and $6,000 for materials to build their own home; $10,000 in all. They figure that by tightening the belt, they can save $2,000 per year. In five years, it seems, they'll have their grubstake. Sadly, this is not the case. If they invest their money at 6 percent interest, but the cost of land and building materials escalates at 10 percent, it actually takes *nine* years to put aside the full grubstake.

Four years late and one dollar in the bank. The exact figures here are not as important as the relationship between savings and prices. This situation is almost as bad as being locked into a mortgage! If inflation takes a hyper rise, as many economists fear will happen, savings will take a tremendous dive in terms of spending power. Our poor friends will stand still, at best. At worst, their real savings will be virtually wiped out.

The only way that this hypothetical couple could manage to save the necessary grubstake in five years would be to increase the amount that they are laying aside each month or each year. If they are paying shelter costs already, this may be extremely difficult. Even if their salaries rise, occasional but major expenses such as a replacement vehicle or medical care may cause a setback in the savings strategy.

Roy's "Second Law of Empiric Economics" might be stated: Savings accounts have a certain optimum size, neither too thin nor too fat. A fat savings account is at risk in a volatile money market and is one of the poorest hedges against inflation. After a healthy savings account is established, a better strategy to further dollar accumulation is to invest the monthly savings immediately in the very things required for the house; land payments, tools, building materials. A "healthy" bank account, in this case, is one which allows you to take advantage of genuine bargains when they come up. I feel that two to three thousand dollars is the ballpark at this writing. This is usually enough to place a down payment on a piece of property, for example, when "just the right piece" becomes available.

A nine-year—or even a *five*-year—strategy has another built-in problem which negates its value for all but the most patient and methodical. The initial enthusiasm for the project diminishes. It's easy to do things when we have to; self-discipline is very much harder. We are creatures of habit and our economic servitude begins to take on the guise of security. Another danger: one of the partners in a marriage gets impatient with far-in-the-future schemes and castles in the air under which no foundations seem to be forthcoming, putting a strain on the relationship.

But there is a good chance you will not be "starting from scratch" in an economic sense. By the time most people have reached the prime age of the first-time home buyer, usually about age 25 to 34, a certain amount of equity has been accumulated. This equity may be in stocks or bonds, a bank account, cars, appliances or some other investment; perhaps there is a modest inheritance. Many people reaching the age of 30 may be surprised to find that their estate has grown to a value of $5,000 or $10,000. Sadly, others attain this age only to find that they have accumulated little of liquid value. If this is you, don't despair! Read on. And take comfort that tomorrow really *is* the first day of the rest of your life.

As a rule of thumb, anyone who can afford to buy a contractor-built home with a down payment and the assumption of a very expensive mortgage, can certainly afford to build his or her own home at a great savings.

Jaki and I were fortunate in being able to start our homesteading venture with the full grubstake. Our grubstake came mostly from the sale of the old stone cottage in the Scottish Highlands, which I'd bought in 1969 for $2,400, renovated (sweat equity), and sold in 1974 for $26,000. This enabled us to move to the United States, buy the land and a small pick-up truck, build the Log End Cottage, and cover our living costs while building. Our income during those first three years—earned mostly by casual labor—was less than $3,000 on average, but that wasn't bad, as our cost of living (above building costs) was not much over $4,000. The grubstake took care of the difference.

To summarize on the full grubstake strategy: If you've got it, use it. If not, that's fine, too. The land grubstake, the temporary shelter, the pay-as-you-go house and other strategies are all open to you.

The Land Grubstake

This is the most common route to achieve the desired goal of freedom from outrageous shelter costs. Most of the case histories later in the book illustrate this strategy. In brief, the idea is to save the money for the land (or at least a substantial down payment thereof), to move onto the land by one of the various temporary shelter strategies (saving interim housing costs) and then to build the permanent house on a pay-as-you-go basis. The total cycle from starting with zero savings to owning your own land and home can range from three to six years, depending on your industriousness and the lavishness of your house and land plans.

One couple living near us, started with about $1,000 and an irregular income—were living debt-free in their own home on a 20-acre woodlot within a year! The home was not completely finished in that time, but it was comfortable even in the north country winter. Now they are finishing off the house.

Because of the favorable price obtained through collective purchase, and the fact that the seller did not wish to take in too much money the first year for tax reasons, many of the people who originally joined in with us on the hill in 1975 were able to buy their land for a down payment of $600. The value of these two strategies—collective buying and securing owner-financing —cannot be overemphasized. Each individual family obtained a Warranty Deed (the best kind) for their parcel. Acreages varied from 15 to 23, selling prices from $2,800 to $3,600. The owner carried the mortgage for five years. Land payments were made once a year, one-fifth of the principal each year plus interest at 6 percent on the unpaid balance. This meant that each year's land payment was a little less than the previous year's. Combined with the decreasing value of money through inflation, this is a very favorable situation for the buyer. Of course, 6 percent mortgages are about as common as nickel beers nowadays, but in 1980, financing through the seller could still be obtained at a rate of 8 percent in our area, well below the national home mortgage average of 13½ percent.

All of the parties who joined together in 1975 to buy land on the hill, and at least three families who have moved to riparian land since, own their own homes and land now, free of debt. That the people are happy with the results of their efforts is shown by the fact that no family has left the hill since starting out, a remarkable record for a community drawn together as haphazardly as ours.

Buying land on a mortgage may *seem* almost the same as buying a house on a mortgage. Yes, there are similarities in kind, but there is a vast difference in degree. A $70,000 house (see article reprinted in the Preface) "is just a shade above the average price for an existing home. With a 20 percent down payment and a 30-year loan at 13½ percent interest rate, annual mortgage payments come to about $7,700." The total price paid over 30 years is: $14,000 plus ($7,700 times 30 years) equals $245,000. A piece of land costing $5,000, with terms of $1,000 down and four annual payments of $1,000 plus 10 percent of the unpaid balance costs $6,000 over the four years: $1,000 d.p. + $1,400 + $1,300 + $1,200 + $1,100.

The cost of land is a necessary evil. And some people, products of the buy-now-pay-later generation, are only able to lay money aside to meet real debts. The self-imposed kind of discipline needed to accumulate savings is just too difficult. For these people, making a few land payments and building equity is certainly preferable to blowing the money away, the inevitable alternative.

Okay, but we've still got to start with something, even if it's only $600 for a down payment on the land, $300 for a temporary shelter, and another $300 or $400 for living expenses until income starts to flow again. Plenty of young people with courage all over this country start out with $1,000 and less, but remember that a proper balance between time and money should be the goal in determining the amount of the grubstake. Low-budget living *can* be a spiritual experience, but just as often, money worries can be a disturbing cause of friction within the family.

How is this paradox resolved? And how much is the right amount for the grubstake? Sadly, there are no hard and fast answers to these questions, as the answers will vary from individual to individual, couple to couple. There are people who think of themselves as "hard up," but would consider themselves *destitute* if forced to live the way of the Roy family: hand-pumping every drop of water used in the home, living without 115-volt A.C. power, flushing the toilet with a bucket. What would they think of using an outhouse in sub-zero temperatures, having no electricity of any kind and hauling the water from a distant pump in buckets, as some of the hill folk do? And yet it would not occur to a single person on the hill that he or she is *poor*. On the contrary, I suspect that, upon reflection, most of us would say we have real *wealth:* fresh air, peace and quiet, good neighbors, clean water, untainted vegetables and lack of debt; in short, *real*—not *paper*—security. Any one of us could trade in our sweat equities for a profit in the dollar sense and readopt the ways of the "masses of men." I, for one, am not so inclined. A man who has once known freedom makes a poor slave compared with the man who has known no alternative to his servitude.

So much for the spiritual side of poverty. The resolution to the paradox might be that we are not speaking of true poverty at all, only dollar poverty and lack of the "gold and silver fetters."

Even though Jaki and I started our homestead with the full grubstake, we were each well acquainted with dollar poverty before we began our partnership, and, as a kind of refresher course, we returned to that vantage point after the completion of Log End Cave. We were broke. And it wasn't much fun. Like Siddhartha (*Siddhartha* by Hermann Hesse), after years in the materialistic world, we'd forgotten how to be at peace with dollar poverty. We buckled down to create that necessary transfer from enforced poverty, in which piddling dollar concerns overwhelmed us, to voluntary poverty, in which we'd have enough to meet our needs. We were fortunate that our relationship was strong enough to take us through a trying period.

If the move to the land can be accomplished via the temporary shelter, and regular income from employment continues without interruption, only the materials cost of the temporary shelter *needs* to be grubstaked. Monies formerly delegated toward shelter costs, usually rent, now can be put directly into building materials. The problem here is often one of an insufficient "stockpile" of funds to maintain the *momentum* of building. Time spent waiting for enough money for the next stage of construction is frustrating. Preferably, enough additional money should be available so that you can make a substantial start on the permanent dwelling, holding down the time you have to spend in the temporary shelter to a minimum. A year in cramped quarters is quite enough for anyone, and especially for a family. If a sufficient grubstake of $2,000 to $5,000 is saved (depending on the house), the period spent in your temporary shelter can be reduced to the six warmest months, when being outside provides pleasant escape from the cramped conditions.

If your regular income terminates with the move to the temporary shelter, you'd better have enough to build the permanent home *and* for at least six months' worth of life's normal expenses during construction. Combine the estimates for these items and add 25 percent as a safety margin. For example, you may estimate that your house will require $6,000 and take six months to build. Figure to spend $80 a week over and above building costs, or $2,080, for a combined total of $8,080. The 25 percent safety margin of $2,020 brings the grand total to $10,100. If land payments are due during construction or shortly thereafter, they must be added.

(Are you shocked at these figures, finding them unreasonably low? "What kind of house can be built for $6,000?" you may ask. The answer is: a comfortable, debt-free home, with a minimum of frills. Or do you despair at the thought of saving $10,000 plus? Don't! It can be done for a lot less, as illustrated by some of the case histories later in this book. The point is that

$10,000—or $5,000 or $15,000—is better invested in building your own home than in a down payment on a bought home or in spending the equivalent on two or three or four years' rent.)

The problems in determining how much of a grubstake to start with stem from four variables: (1) Everybody's different. Their abilities are different. Their fiscal responsibility and earning potential varies. Their house and land requirements are different. (2) Land prices vary tremendously around the country. (3) Earnings and cost of living vary around the country. (4) Building and planning codes vary from overly restrictive to nonexistent. (The chart which can be found in the Appendix shows what strategies are possible with different combinations of grubstakes and earning power.)

Even if the temporary-shelter strategy isn't employed, the land grubstake is still a viable strategy. Your savings are better invested in land, which tends to rise in value at least as fast as inflation. Land payments have a static price and their interest rises at a slower rate than that of inflation.

The "roughing it" implied with the temporary shelter may not appeal to you or your partner, and is not imperative as long as you buy the land outright or on a mortgage. If further savings are converted directly into building materials—which can be stored on the land providing that they are safe from theft and deterioration—the problem of escalating building costs is also kept under control. Converting indigenous raw material on the land—standing trees into logs and log-ends, for example—and recycling of building materials such as are found in old barns in the area, can bring the building reality in sight, perhaps within a year or two. A well can be dug and capped, a septic system installed, a driveway built, all at current prices and not at some future inflated price.

Lots of other valuable but not expensive steps can be taken in the interim period between the purchase of the land and construction of the house. Susan Warford's list of her homestead-related enterprises conducted while living hundreds of miles from her land is remarkably complete, and she shares it with us in a later part of this book.

Perhaps you already live on a family farm, or your parents own country land which they'd be willing to share with you. Great. Count your blessings. Now the only part of the grubstake that needs to be considered is construction materials. However, be sure to get and record a deed for the property prior to any construction. *"But Mom and Dad wouldn't cheat their dear and loving only son!"* No, no, of course not. (Although I know of such a case.) But there are two points to keep in mind here. One is that, just as in the case of friends joining together on a land purchase, there is more at stake than just land or money or houses, so it is imperative that there be no misinterpretation of your agreement. Sometimes these family "understandings" are far too casual, and, after a few years, each party to the understanding remembers it just that wee bit differently. A family feud can then erupt.

The other potential problem that could arise is when a probate court gets involved in the family feud. Probate courts are concerned only with *facts,* like *written* agreements and *recorded* deeds. Do it right. Never, under *any* circumstance, build a house, dig a well, or conduct any valuable improvements to property *which you do not own by clear recorded title.* No one can foresee all the circumstances which can foul up a casual agreement. Based on my personal experience and observations of the experiences of others, the success ratio of nonbinding agreements is less than 50 percent. Buck the odds at your own risk.

How to Save the Grubstake

Barring an unexpected inheritance or sweepstakes prize, it will be necessary to lay up your own grubstake. The rate at which this money can be saved is proportional to the degree of austerity which you can accept. In other words, exactly how much are you willing to do without in order to reach your goal in the shortest possible time?

To answer the question you must be able to differentiate between necessities and luxuries. Let's examine the necessaries more closely, to try to determine the difference between that which is truly a "necessary of life" and that which is not.

FOOD

Without going into a detailed nutritional analysis, it can be safely stated that the more common form of malnutrition among American adults is obesity, not undernourishment. When youthful activity decreases, as early as the mid-20s for many people, and the habitual diet is maintained, body weight will increase. The alternatives are increased exercise (building a house?) or a decreased caloric intake. Just as forest management can yield firewood *and* an improved woodlot, a change in diet can yield dollar savings *and* improved health. Americans, for example, consume an inordinate amount of meat, especially fatty meats such as streaky bacon and beefsteak. This is good neither for national nor bodily health and, as protein production, is a wasteful use of agricultural lands. According to agricultural scientist David Pimentel of Cornell University, Americans consume two and one-half times the amount of protein recommended by the United Nations Food and Agricultural Organization. To accomplish this, we grow 10 times as much grain to feed cattle and other livestock as we consume ourselves.[19]

Jaki and I are not vegetarians, nor are we advocates of vegetarianism for any lofty spiritual reasons, although we respect the legitimate views of others on this. But we buy very little meat at the market, although we are inclined to order meat when eating out. Aside from health benefits, decreased meat consumption gets the saver into an eating pattern which is easy to maintain by growing food inexpensively in the garden.

Eating out is not food consumption for survival, but rather for entertainment. We engage in this luxury more now that our house is paid for. During times when it was necessary to save money for some purpose, dining out was one of the first items to be scratched from the budget.

FUEL

The standard suggestions, increasingly heard now even in the mass media, really do work: car pooling, organizing chores to avoid duplication of trips and wearing a sweater so that you can turn down the thermostat to 68° F. Rooms in the home which are seldom used need not be kept heated while not in use. Gas and electric hot water heaters should only be flipped on if a significant amount of hot water is required; otherwise, heating a small quantity of water or keeping a kettle on the woodstove is more economical. Plan hot water usage for specific times of need.

CLOTHING

Clothing expenses can offer a tremendous opportunity for savings by eliminating fashion considerations and concentrating instead on insulating the body. We have found thrift shops and Salvation Army stores to be excellent sources for clothes, and we even sew on patches at worn but "strategic" locations.

SHELTER

The big gain here will come later, which is the point to this book. In the meantime, be alert to opportunities for cheaper shelter costs—a lower rent situation, perhaps taking on a house caretaker's position or sharing your accommodations, chores and expenses. When collective land purchasing is an anticipated strategy, the time spent together in close quarters during grubstaking might help to confirm the group's compatibility. If you can survive this intact, you've got a strong partnership of great value when the move to the land is made.

LUXURIES

As has been seen, there is room to save, even where the necessities of life are concerned, but the real impact will come from cutting back on the luxuries. A friend of mine smoked three packs of king-size cigarettes a day while complaining of being unable to afford many things. At the time, we figured out that he was spending $2.25 a day for his habit. This amounts to *$821 per year,* a figure greater than either the down payment or the annual land payment (*plus* property taxes) for most of the land parcels in our neighborhood. Need I say more?

Beer is my vice. When we're short of cash, Jaki and I brew our own beer, at a tremendous saving. As with most of the savings tips in this section, the gain is not only in dollars, but in quality. (You wouldn't put anti-freeze in your homemade beer, even though it is present in many of the national brands.) Now that we are not so pressed financially, however, we have stopped making our own beer. This is pure laziness on our part and is a big mistake, economics notwithstanding, because the quality of our beer has declined.

Expensive cars and expensive toys—like snowmobiles, large motorcycles and fancy stereos—make grubstaking difficult. If you own any of these machines, appliances or toys and find that they mean less to you than owning your own home, sell them now. Unlike economy cars and practical appliances, like a cast-iron cooking range or a fine set of tools, luxury items depreciate very rapidly. If the resale value seems low today compared to the original price, rest assured that it will be worse tomorrow.

It could be argued that entertainment is a necessity of life. You know what is said about people who are "all work and no play." However, you may not have considered the many inexpensive—and even free—entertainment activities and social events that are available. Some can offer the opportunity to make new friends and learn new skills. Instead of going to expensive rock concerts or buying record albums, why not learn to make music through local music clubs? The publicly-owned television or radio station in your area may provide more satisfying entertainment than a disappointing but high-priced movie. Or get back into reading. You don't have to buy the books; that's what libraries are for. Learn more about securing your economic freedom by reading some of the fine works mentioned in this book or currently on the library shelves. Later, when you're really ready to start building, you'll probably want to own those books which you have discovered that best explore and explain your choice of building or life-style.

I realize that some of the money-saving tips offered above might involve a departure from your current value system. But, I also know that when you desperately want to build your own home in the country you find a way to tighten your belt and make the necessary sacrifices. Most people who have already made the "sacrifices" report that they really weren't sacrifices at all, but were, in fact, qualitative improvements in their physical and spiritual well-being.

In addition to everything else, austerity budgeting is an education. The skills and self-discipline learned will help guide the way when economic impoverishment is *not* voluntary, particularly during house construction.

Whole Earth editor Stewart Brand's outlook on economics is almost Zen-like in its wisdom: "Living below your means is a cheap way to be rich. It's the only way to be rich."[20]

And when you get really sick of skimping and saving and watching every penny, there is no sybaritic pleasure to compare with giving in and indulging in a steak or lobster dinner with a bottle of fine wine.

4

The Temporary Shelter

Obtain the land at the earliest possible date. It's the best hedge against rapidly escalating real estate prices, and protects your hard-earned dollar from devaluation. Equally important, land acquisition immediately accelerates your rate of savings—perhaps by 100 percent—so that the house grubstake can be accumulated much faster. The key is the temporary shelter.

The Strategy

In its classic form, the temporary shelter plan involves building a small, livable, low-cost structure on the land, and living in it while you build your permanent house. There can be variations on this strategy, as will be seen, but its potential advantages essentially are these: the gaining of building experience, the elimination of interim shelter costs, the gaining of knowledge about the land itself and the gaining of a useful outbuilding for extended use.

THE GAINING OF BUILDING EXPERIENCE

Most owner/builders start with a common background of inexperience in construction, which, of itself, is not really that great a handicap. Unfortunately, many would-be builders are *fearful* about their lack of experience, and this can evolve into a lack of *confidence,* which *is* a problem. But, by methodical application of building techniques to a small project based on a simple set of plans, soon you will find that, *I really can build a structure in which I could survive.* This is a giant step, psychologically. In a month or less, you will move from dependency on others for your shelter to full equality with birds, bees and beavers! Jaki and I fondly remember our first days in our temporary shelter, a 12-foot (3.66 m) by 16-foot (4.88 m) shed, and the sense of accomplishment which has not been topped by any of our subsequent building projects. Although I'd acted as my own general contractor for the cottage renovation in Scotland, my actual building experience was virtually nil. I'd done a lot of the destructive work, but the skilled carpentry, masonry, plumbing and electrical work was sub-contracted to skilled tradesmen.

My late father designed two houses and had them built by professional contractors. "You have to build two houses to get one right," he would say. What he meant is that you get the mistakes out of your system with the first house—90 percent of them, anyway.

My experience is that some mistakes will continue to be made as long as new techniques are tried. My father also used to say, "We should learn from our mistakes, of course, but the wise man learns from the mistakes of others." If you want an error-free home, buy a set of building plans for a simple framed structure which has been built hundreds of times without problem. Lumberyards have stacks of such plans. The result will be a well-built, somewhat functional, antiseptic home. Sometimes these plans can be adapted fairly easily to the use of rough-cut or recycled lumber, which will introduce a little more scope for individuality into the home.

My father may have been right, but who else except a confirmed building addict like your author is willing to build more than one house? Hardly anyone, which is why the temporary shelter strategy is being introduced.

Now, you should give some thought to the final house, even during construction of the temporary shelter. Practice the building techniques which will be employed on the main structure. Why build a cordwood masonry temporary shelter if a traditional log house is the desired permanent home? Your temporary shelter can be of post and beam, adobe, stone, underground, timber-framed, cordwood, log or A-frame construction—or any other type which is desired for the main house. While *any* building experience is useful, the full advantage of the strategy incorporates practice of the techniques to be used in the future. This is particularly important architecturally if the temporary shelter is to be recycled as a part of the main structure later on.

THE ELIMINATION OF INTERIM SHELTER COSTS

Usually, the temporary shelter can be built in from two to four weeks, or a little longer if work is limited to weekends and spare time only. After that, the move to the land can be made and the portion of the budget formerly devoted to rent or mortgage now can go toward materials for the new home, minus any increase in commuting costs resulting from the move. Even if the full grubstake is already laid up, the temporary shelter provides rent-free accommodation during construction. In this case, the elimination of the additional cost of traveling to the building site is a further savings, a safety margin.

The convenience of living on site during the permanent house construction cannot be over-emphasized. We enjoyed this advantage during the building of Log End Cottage and again at Log End Cave. If you fancy working for an hour after supper, you're there to do it. If you have to drive 10 miles (16.09 km) to the site, time and money are lost and you're less likely to make the effort.

THE GAINING OF KNOWLEDGE ABOUT THE LAND

So often people decide where to build their house after one or two visits to the land, even as to the exact alignment of the structure. A certain view, the location of the road, or some other seemingly obvious factor make other considerations inconsequential. Frequently, the choice involves romantic, rather than practical, decision-making. There's nothing wrong with romance, but it lasts longer if coupled with "stuff" of substance, like respect for sun and wind direction, ease of construction (can a concrete truck get to the site?) and integration with the other living systems (to be discussed in Chapter 5). These considerations will become much clearer to you after even a couple of weeks of actually living on the land.

Bob Easton, writing in *Shelter* (Random House, 1973) says:

> It would be ideal to camp out on your site for a year and watch the changes before building anything. An alternative would be to build a small shed to the side of your site and live there for a year before deciding what to do next. You could watch the angle of the sun change throughout the year, learn where the winter storms come from, and figure out how to have the morning sun at your breakfast table. You could see how to catch cool breezes in the summer and see the stars at night. Also, you'd have time to meet the neighbors, study their houses, talk to the old folks about elements peculiar to the area: special winds, drainage problems, sources of cheap materials: local wisdom about local problems.[21]

Mr. Easton's advice is sound, if you're of the frame of mind to take advantage of it. For many, a year in a small shed prior to starting the house might be a bit too austere. We spent eight months in our shed, and although we were sad to leave the first home we'd actually built ourselves, it sure was a pleasure to have a few hundred more square feet in which to move around, especially as a north country winter was about to heap its fury upon us. We have known other families whose time spent in the temporary shelter was just too long—two years and more, for example—and the initial high of moving to the land turned to dissatisfaction with the cramped living conditions. Apart from everything else, it is very difficult to maintain a steady temperature in a small structure in the winter. The place oscillates from sauna to freezer-like conditions.

THE GAINING OF A USEFUL OUTBUILDING

The buildings discussed here may be temporary *shelters* insofar as human habitation is concerned or they may be incorporated into the final house plans —but they are certainly *not* short-lived structures. (An exception is the

12 × 16 GABLE
(not to scale)

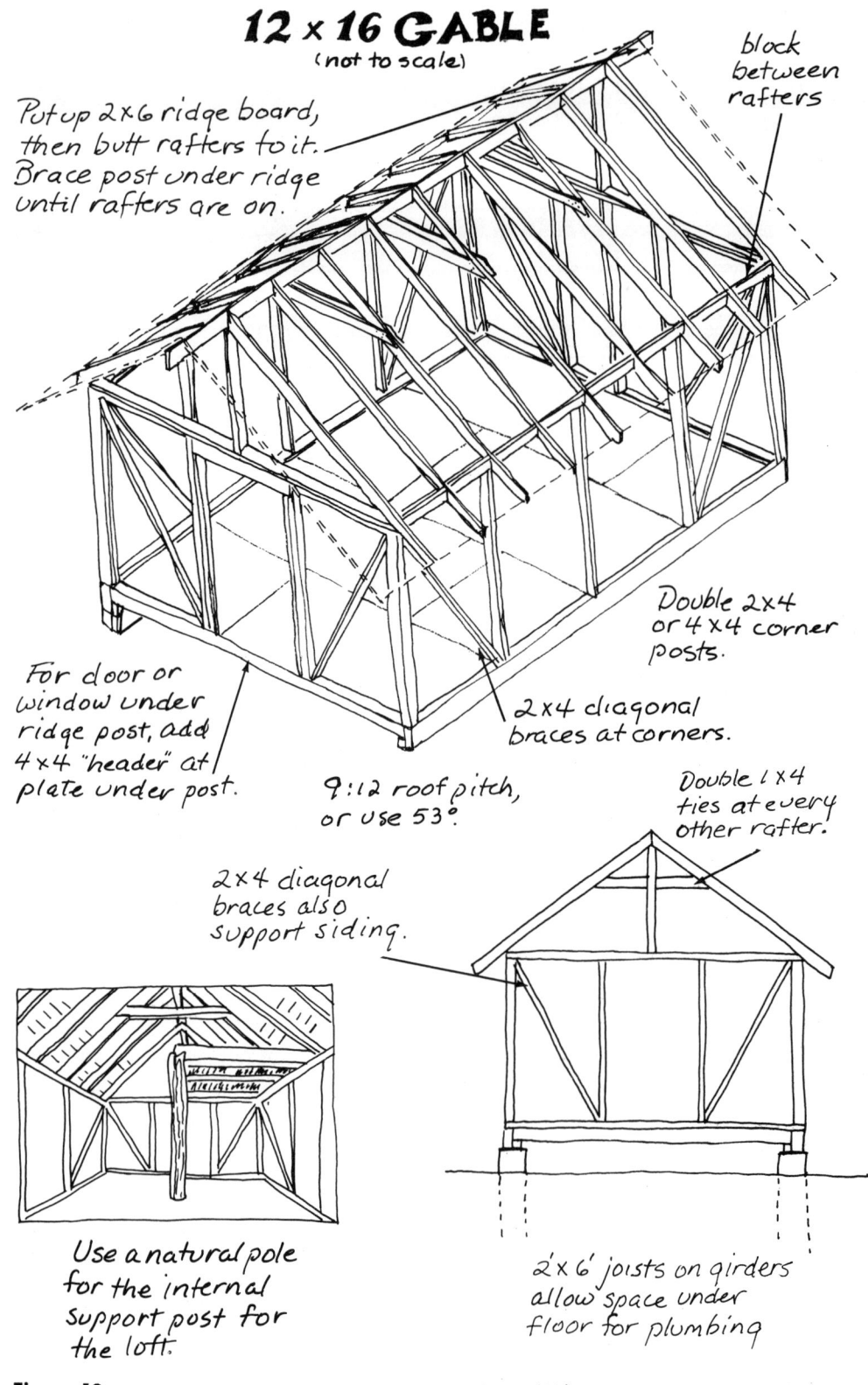

Put up 2×6 ridge board, then butt rafters to it. Brace post under ridge until rafters are on.

block between rafters

Double 2×4 or 4×4 corner posts.

For door or window under ridge post, add 4×4 "header" at plate under post.

2×4 diagonal braces at corners.

9:12 roof pitch, or use 53°.

Double 1×4 ties at every other rafter.

2×4 diagonal braces also support siding.

Use a natural pole for the internal support post for the loft.

2'×6' joists on girders allow space under floor for plumbing

Figure 10.

CUTAWAY THRU GABLE AND SHED

temporary cordwood house, discussed later in this chapter, a truly *temporary* structure.)

Some thought should be given to that seemingly far-distant day when the move is made to the permanent house and the temporary shelter becomes empty. It may have 40 useful years of life in it yet. It might function as a garden or wood shed, workshop, playhouse, studio, sauna, guest house, chicken coop, barn or even root cellar, if built underground. The building's anticipated future use is just as important as its temporary job of providing human shelter so you should tailor siting considerations and structural details to the long-term uses.

Scandinavian immigrants settling in the midwest during the last century often adopted the strategy of building their sauna first—such were their priorities—and living in it until the house was finished. It remained as the sauna in later years.

Another common strategy—involving a different set of priorities—is to build the garage first, and to live in it until the house is completed. I have heard of one instance where local codes forbade the temporary shelter strategy. The family moved into their owner-built garage and that was okay as long as they did not have a "kitchen," defined by "permanent cooking facilities." A Coleman gas cooker was trotted out each time they wanted a meal, and that apparently satisfied the building inspector.

Individual Strategies

TENTS, CAMPERS, VANS, AND TRAILERS

Temporary shelters are not limited to built structures. You can live several months in tents, tepees, vans and small house trailers while building your permanent dwelling. In fact, Jaki and I lived in a small trailer while building our shed, a truly *temporary* temporary shelter. In certain circumstances, the time gain and economic advantages of staying in a van or a tent might make this the correct strategy, especially if you already have building experience.

THE SHED

There is no doubt that the common shed or one of its variations is the building most often used with the temporary shelter strategy, and it is particularly recommended if standard framed construction is desired in the final home. We built the 12-foot (3.66 m) by 16-foot (4.88 m) gabled shed (Figure 10) drawn by Bob Easton that appears on page 41 of *Shelter*. A drawing representing our building is reproduced here. Bob Easton's drawing was detailed enough to supply us with the information required to build, yet simple enough not to overwhelm us.

We built our shed on 6 cedar posts set about 3 feet (.9 m) into the ground, instead of the concrete piers shown in the drawing. We cut the tops of the posts to the same level and installed a 2-by-10 girder along each side of the structure. We fastened a 2-by-4 (5.08 cm by 10.16 cm) nailer alongside the girder as shown in Figure 11, this to support the 2-by-6 floor joists. The 13 floor joists were placed 16 inches "on center" (16″ o.c.), which means that the spacing from the center of one joist to the center of its neighbor is 16 inches. The shed was floored with ½-inch exterior grade plywood so the 16-inch joist spacing worked in well with the 48-inch by 96-inch plywood, and was frequent enough so that the plywood didn't sag, as it would have done on 24-inch centers. As it was, our floor had a little spring to it, because we used only 6 posts instead of the 9 recommended by Bob Easton.

Later, when we installed the internal center post to help support the loft (Figure 10, lower left), we had to jam under a couple of large flat stones midway along the center joist for additional support. A better method of using 2-by-6 joists on the 12-feet by 16-feet shed is to use 3 girders as shown in Figure 12a, but this requires 3 additional posts and an extra girder.

Another method of building a strong 12-feet by 16-feet deck involves the use of 4-by-8 girders and joists and 2-by-6 tongue-and-groove planking. (Figure 12b) Four-by-eights can be cut at a small local sawmill for not much over the cost of finished 2-by-8's at building yards. Tongue-and-groove planking is quite expensive, unless the local sawmill has planer knives which will shape this sort of material. Several people on the hill—including ourselves—have been successful in buying up old wooden silos which were rotting out at the base, but still had lots of good tongue-and-groove planking above. The cost of these recycled silo staves—often spruce or cedar—can be less than a third of comparable new material. And keep the silo hoops! They will make excellent reinforcing bars for the footings of the permanent house, saving another hundred dollars or so.

Old Tom, from whom we'd bought our land, showed us how to frame the first wall. He was certainly qualified to teach us, having built 90 houses in one nearby village alone. He also made 3 sash windows for us, bartered for an old hand-cranked sharpening stone which we'd picked up at an auction for $3.00. Steve Dorresteyn, one of the people who joined in on the land purchase, shared his roof framing experience with us.

As we were fresh from Scotland and really needed a place to stay in a hurry, we bought finished studding and new plywood, both from a discount yard. Scrounging around for bargains on materials can save a lot of money, but it does take time. On the Log End Cottage, we did more scrounging and used a lot more recycled materials. Still, we built the shed in three weeks for $350, which wasn't too bad for beginners, we thought, although Steve or Tom certainly could have accomplished the same in three days.

One interesting innovation which we employed successfully on our shed is worth mentioning, as it could save you the price of this book 10 times over if you use the idea on your own house or shed. Our roof is composed of aluminum offset printing plates, like the ones used in the printing of your local paper. These heavy-gauge plates are used once and thrown away or sold for scrap. (Some papers have started selling them to the public, but it's still

Figure 11a. Cedar posts (or pressure-treated timbers) are set 3 feet in the ground with gravel and jamming stones at their bases. The girder and nailer are fastened together and toenailed as a unit to the posts, ready for floor joists. Check all dimensions, including for square, by measuring the diagonals of the 12-foot by 16-foot rectangle.

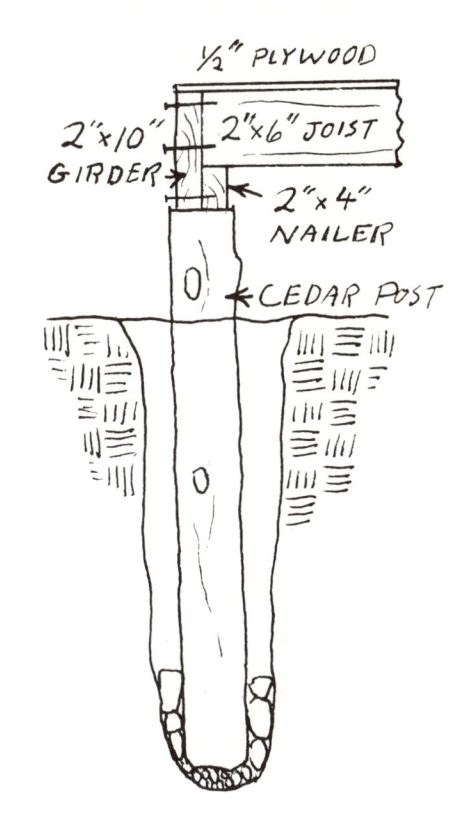

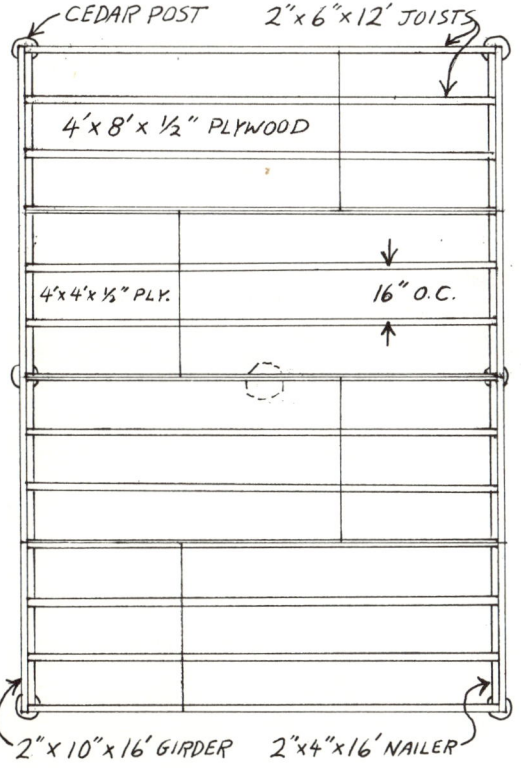

Figure 11b.
Location of the 13 2-by-6 floor joists and the 6 sheets of plywood. (Two sheets are cut exactly in half to provide for staggered jointing.) The circle in the center shows where flat stones were jammed in below the center joist to provide support for the loft post.

WOOD FLOOR FRAMING

There are 2 simple wood floor systems which you can use on the pier foundation. A joist and girder frame with ¾" flooring, or a girder frame with 2x6 T&G.

Joist and Girder
—2x6 joists spaced at 16" on center can span 6 feet. 2x8 can span 10'. 2x12 up to 16'.
--4x6 (or double 2x6's nailed together) girders spaced 6 feet apart to support 2x6's. This girder can span 6 feet between piers. 4x8 can span 10', 4x12 up to 16'.

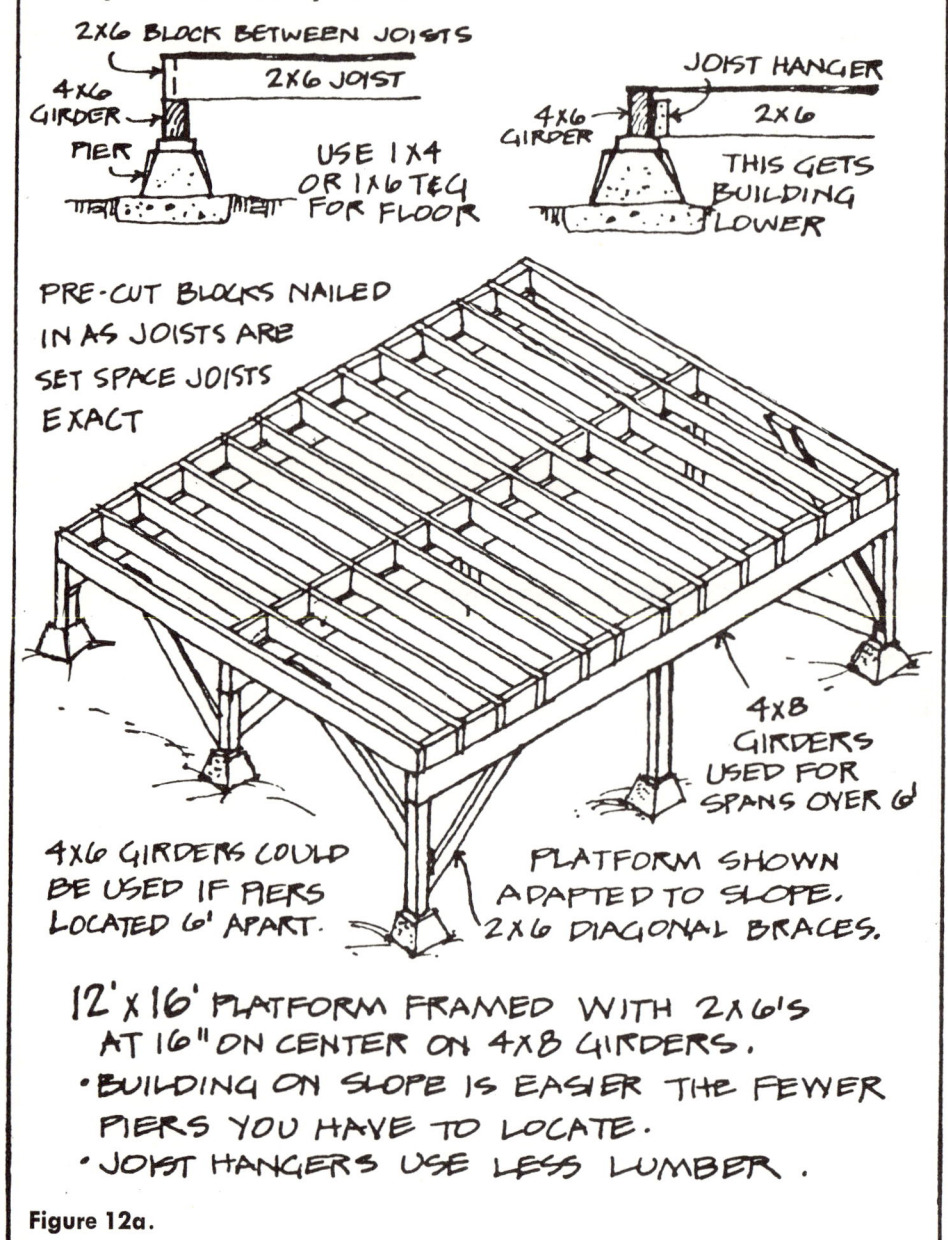

2X6 BLOCK BETWEEN JOISTS
2X6 JOIST
4X6 GIRDER
PIER
USE 1X4 OR 1X6 T&G FOR FLOOR

JOIST HANGER
4X6 GIRDER
2X6
THIS GETS BUILDING LOWER

PRE-CUT BLOCKS NAILED IN AS JOISTS ARE SET SPACE JOISTS EXACT

4X8 GIRDERS USED FOR SPANS OVER 6'

4X6 GIRDERS COULD BE USED IF PIERS LOCATED 6' APART.

PLATFORM SHOWN ADAPTED TO SLOPE. 2X6 DIAGONAL BRACES.

12'x16' PLATFORM FRAMED WITH 2x6'S AT 16" ON CENTER ON 4x8 GIRDERS.
• BUILDING ON SLOPE IS EASIER THE FEWER PIERS YOU HAVE TO LOCATE.
• JOIST HANGERS USE LESS LUMBER.

Figure 12a.

Girder and 2x6 T&G
—4x6 girders spaced at 4 feet on center can span 6 feet. 4x8 can span 12 feet, 4x12 up to 18 feet. Block between girders every 10 feet.
—2x6 T&G can span 4 feet.

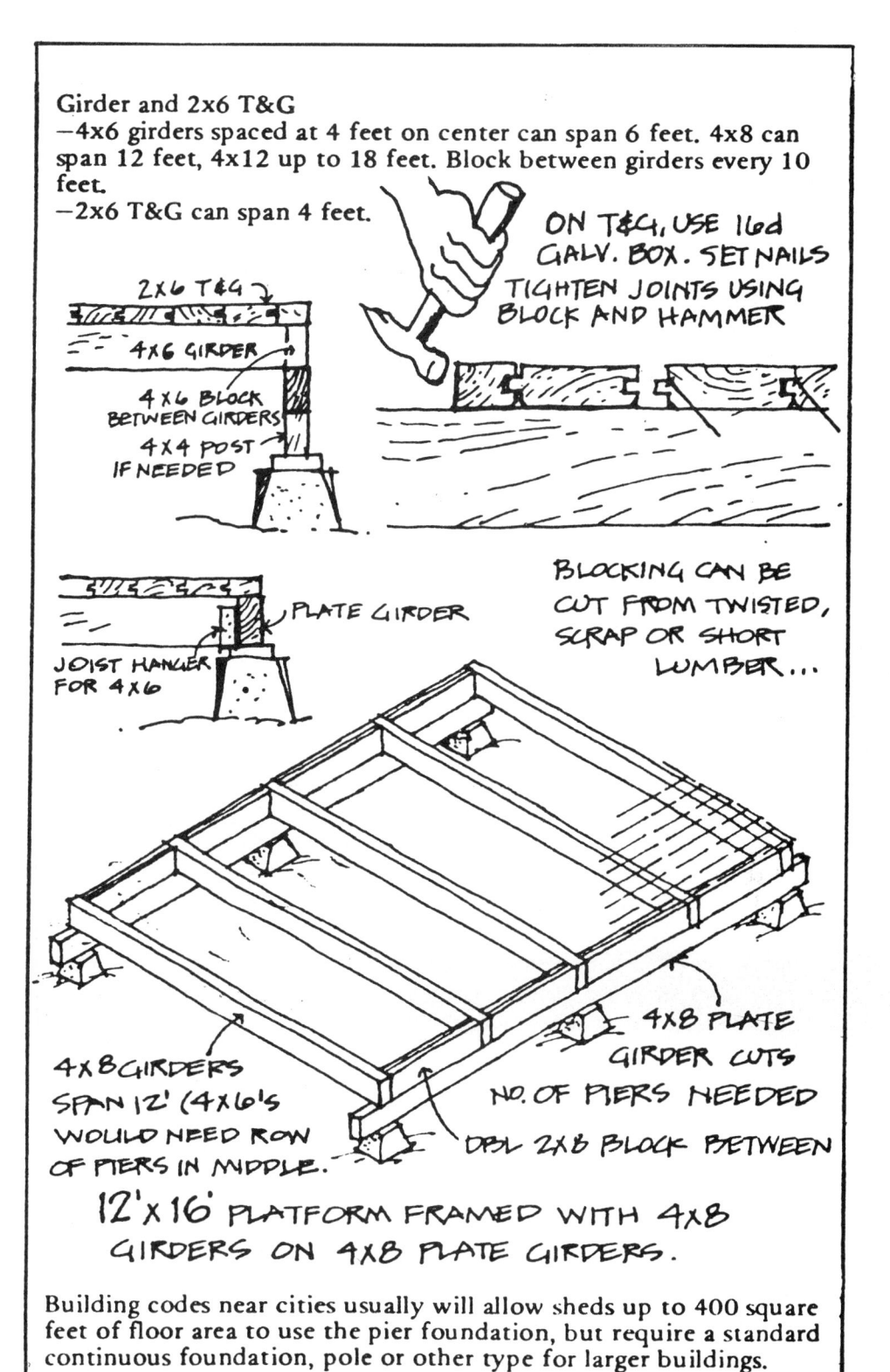

ON T&G, USE 16d GALV. BOX. SET NAILS TIGHTEN JOINTS USING BLOCK AND HAMMER

2X6 T&G
4X6 GIRDER
4X6 BLOCK BETWEEN GIRDERS
4X4 POST IF NEEDED

PLATE GIRDER
JOIST HANGER FOR 4X6

BLOCKING CAN BE CUT FROM TWISTED, SCRAP OR SHORT LUMBER...

4X8 GIRDERS SPAN 12' (4X6's WOULD NEED ROW OF PIERS IN MIDDLE.

4X8 PLATE GIRDER CUTS NO. OF PIERS NEEDED
DBL 2X8 BLOCK BETWEEN

12' x 16' PLATFORM FRAMED WITH 4X8 GIRDERS ON 4X8 PLATE GIRDERS.

Building codes near cities usually will allow sheds up to 400 square feet of floor area to use the pier foundation, but require a standard continuous foundation, pole or other type for larger buildings.

Figure 12b.

possible to get them for nothing or next to nothing if you ask around print shops.)

We left the printed part exposed to the weather, so the roof looks like it is made of newspaper! The plates, nearly 2 feet by 3 feet, go on very quickly, but be sure to use aluminum nails with neoprene (synthetic rubber) washers. Ordinary roofing nails are likely to cause leaking and there is the added danger of electrolysis within the different metals eating away the aluminum around the nails. Nail the plates at frequent enough intervals so that the wind won't work them loose, or, better yet, use 1-by-2 or 2-by-2 battens to hold the edges down, as we did.

We stapled 15-pound builders' felt on the exterior walls, as a temporary measure to protect the plywood. Later, we applied the cedar slabwood seen in Figure 13. We were given this slabwood when Steve had his cabin logs cut, but our local sawyer charges only $5 per pick-up truckload for it and allows us to rummage through the piles for the best pieces. By all means, make friends with your local sawyer. Even the by-products of lumber milling are valuable. We use the sawdust quite extensively in cordwood masonry, for example.

We added the greenhouse to the shed two years later, giving it its present appearance. The building has proven invaluable as a storage shed, garden shed and greenhouse. During the winter of 1979–80, it even served as a peaceful office while I wrote *Cordwood Masonry Houses*.

Thoreau's house on Walden Pond, incidentally, was 10 feet wide and 15 feet long, smaller than our shed. He built it at a cost of $28.12½ in 1845. "I thus found that the student who wishes for a shelter can obtain one at an expense not greater than the rent which he now pays annually."[22]

Students paid $30 a year for a room at Cambridge College at that time. (It is interesting to note that a kit based on Thoreau's cabin design is now being offered by the archaeologist who discovered the Walden homesite just after World War II. Cost? About $4,000.)[23]

A smaller shed with the traditional single-pitch "shed roof" is shown in Figure 14. This structure is quick, easy, and cheap to build and would supply adequate temporary shelter for a single person. Six pillars or posts would be plenty for this size of shed.

Susan Warford, whose "case history" appears in Chapter 6, felt that the A-frame style would be the easiest for her to tackle (Figure 15). She and her father built the basic shell of her 12-foot by 16-foot shed in a week. Susan moved in. As the weather was mild and Susan did not plan on building the main house for a year or two, she took her time "winterizing" the shed. Although the little place was difficult to keep at a steady temperature, Susan did brave the hard winter of 1978–79 in the A-frame, after a total outlay of $550, including insulation and stove. Most of the materials were bought new. The platform for the building is the same size as our shed's, but the immediately receding walls greatly decreases the useful floor area and the practicality of shelves. Still, Susan found the construction to be quick and easy and recommends it as a short-term shelter.

Figure 13.
This 12-foot by 16-foot shed, with its cedar slab siding and printing plate roof, was our temporary home for eight months while building the Log End Cottage. Later, we added on the greenhouse that you see attached to the main structure here.

8'x 12' SHED
(not to scale)

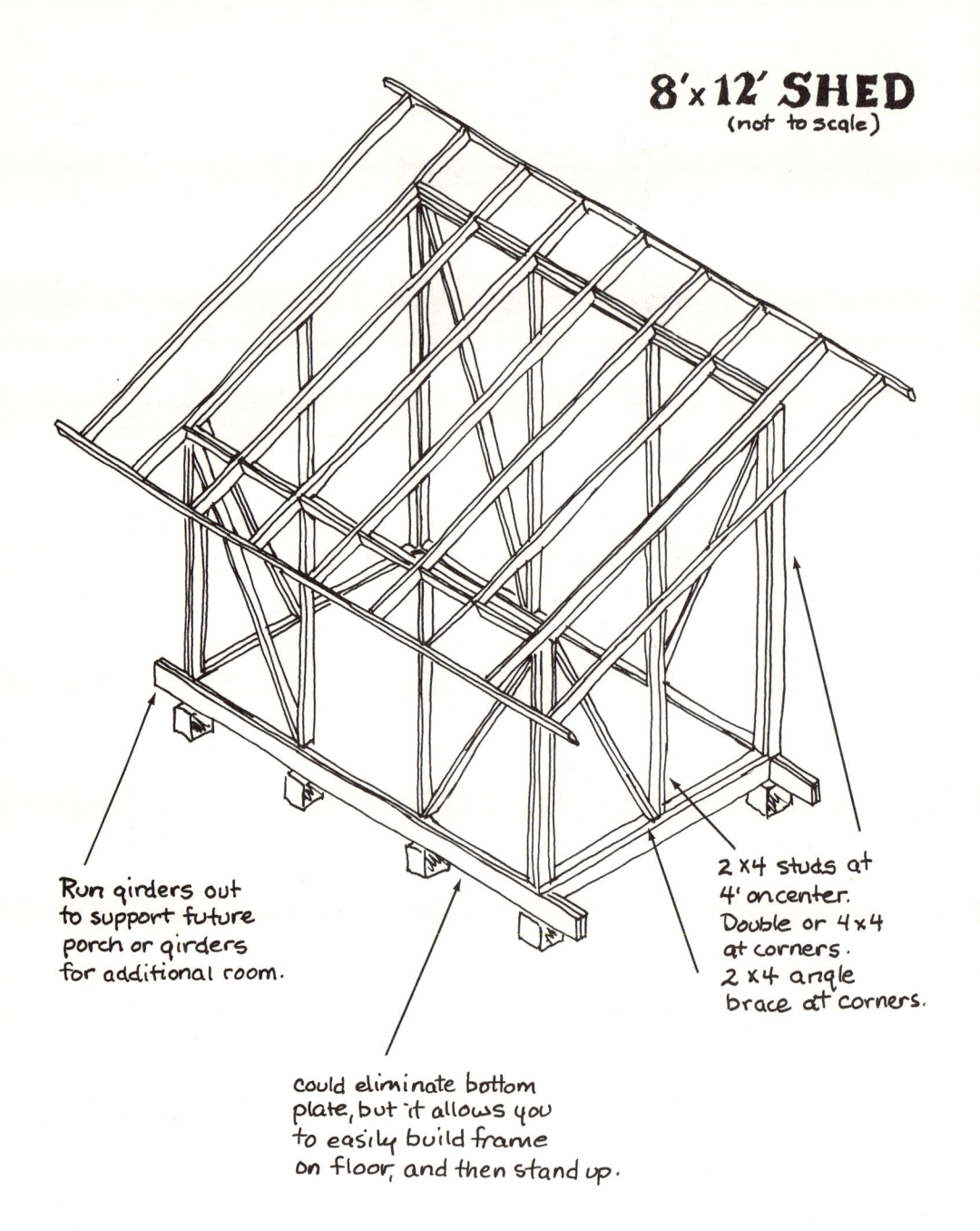

Run girders out to support future porch or girders for additional room.

could eliminate bottom plate, but it allows you to easily build frame on floor, and then stand up.

2 x 4 studs at 4' on center. Double or 4 x 4 at corners. 2 x 4 angle brace at corners.

Figure 14.

Figure 15. Susan Warford, with the help of her dad, built this 12-foot by 16-foot A-frame temporary shelter, then built her capped basement.

Rich and Anne McIntosh's temporary shelter, seen in Figure 16, is the classic case of a building coming full circle in its function. Rich bought a chicken coop from a local farmer in 1975 for $50 and transformed its materials into the barn-shaped structure that became their home for three years while building a large octagonal log house. Today, their temporary shelter has returned to duty as a chicken coop!

Artist Pat Duniho's tree-stump house (Figure 17) started out to be a temporary shelter, but has evolved into something between that and a permanent home, so I will leave the telling of his interesting story for Chapter 6.

CORDWOOD MASONRY IN A TEMPORARY SHELTER

Cordwood masonry, the construction technique in which short logs are laid up widthwise in a wall like a rank of firewood, offers excellent opportunities for a low-cost temporary shelter. The inexpensive New Brunswick cabins of

Gott Selte and Ferny Richard, featured as case histories in my book *Cordwood Masonry Houses* (Sterling), are good examples of this. Gott's small, five-sided structure, with its earth roof, was built in three weeks with hand tools only, at a cost of $180. Ferny's place, a much more substantial building with 395 square feet gross (320 square feet net due to the 12-inch-thick walls), cost only $1,000 for materials and $75 for hired help in 1975. In 1980, Ferny told me that he could build the same structure again for half that money, knowing what he knows now.

The Richard cabin is what I call a second-generation temporary shelter; that is, its size is such that several years could be comfortably spent in such a building, allowing more time for the accumulation of the grubstake and materials for the permanent home. It could even be incorporated into the new home as a unique and beautiful living room or den. In any case, its low cost qualifies it for consideration as a part of the general strategy described in this part, even though 500 man-hours were required to build it (equivalent to two people working five 50-hour weeks).

In *Cordwood Masonry Houses,* I described the construction of a cordwood masonry sauna with outside wall dimensions of roughly 10 feet by 15 feet. I planned this structure very carefully and gave clear instructions on how it could be built, even though *I had not yet built the structure myself.* Before the publication of the book, and using the publisher's galley sheets as my guide, I did in fact build the sauna, following my own instructions practically verbatim. The result is shown in Figure 19, which can be compared to the originally published line drawing (Figure 18). I mention this not to brag that my plans were so good, but because the sauna could be altered slightly to become an excellent low-cost temporary shelter. In point of fact, the actual structure cost us about $600 in materials, of which $125 was for the stove and insulated stovepipe, and $50 for reusable footing forms.

If the building were to be designed as a temporary shelter, I would advise increasing the dimensions to the 12-foot by 16-foot size, which we used on our shed, because considerable floor area is lost in the cordwood walls themselves. Incidentally, the sauna took about 250 man-hours to build, a little longer, perhaps, than a simple framed structure of similar size, but this sauna is built like a fortress and should last a century or more.

One of the problems with cordwood masonry is that very dry wood must be used in buildings intended for long-term living. Otherwise, wood shrinkage necessitates a tedious caulking job a few years down the road. Sometimes wood that is dry enough for permanent cordwood construction just is not available. There are two options open to the you, each involving construction with "green" (unseasoned) wood.

If the temporary shelter's ultimate purpose will be unhampered by slightly drafty walls—a garden shed, wood shed or small barn, for example—there's no reason why a building of green log-ends will not suffice as human shelter for a few months, even a year or two.

The alternative to the temporary *shelter* of cordwood is the temporary

Figure 16. Rich and Anne McIntosh lived in their gambrel-roofed shed while building a log house. The shed, made from a recycled hen house, has now been returned to its original duty with the completion of their log house.

structure, a different strategy altogether. This plan eliminates the last advantage of the temporary shelter—the gaining of a useful outbuilding later on—and replaces it with another advantage: the seasoning of cordwood for future use.

Cordwood Masonry Temporary Structure Strategy

The strategy is this: you buy the land and move on to the property, even if it means camping for a month. You bring with you a good chainsaw (worth spending the money for this item), a trowel, a hammer, an axe, a wheelbarrow, a shovel and old hoe for mixing mortar and lopping shears for easy tree trimming. Other necessary tools can be bought as needed, but the ones listed will hold you in good stead for a long time, so you might as well get good ones. Tools are one of the primary expenses for the temporary cordwood structure, but they will be used again on the permanent dwelling, so it's money well spent.

Armed with good tools, you go out on your woodlot, cut log-ends like a madman, mortar them up (with weak mortar) on the cheapest temporary foundation you can think of, and roof over the building with one-by planking and 30-pound roll roofing (or some other low-cost roofing). You move into the structure and heat it with a woodstove. (If the economics allow, you might as well get a good stove that can be used again when the house is rebuilt.) When the cordwood is dry, you demolish the structure with a sledgehammer and rebuild it on a proper foundation which you've built in the meantime.

Crazy? Not necessarily. Experience is gained in laying up cordwood masonry, the log-ends are dried out at an accelerated rate, thanks to the wood stove's effect in winter, and a move to the land is made which might not have been possible otherwise. *Cordwood Masonry Houses* goes into the detail on this strategy, even illustrating the construction of such a temporary structure by the modular plan: When it is time to build the permanent home, half of the temporary structure is torn down and rebuilt on the permanent foundation. The owners live in the remaining half in the meantime. Then the move is made to the completed half of the new structure. The second half of the original building is razed and rebuilt on the new foundation. The disadvantages are the extra couple of months' work involved in providing your own permanent shelter, and, of course, the hardship involved. The decision is yours.

Not everyone in the United States has access to cordwood, of course. Maybe sod or stone or adobe is the indigenous building material. No problem. The same principles and advantages of the temporary shelter are present.

FROST PROTECTION

In northern climates, where frost heaving is a problem, the foundation is extremely important, especially for masonry construction. Our sauna is built

Figure 17. Pat Duniho's Stump House was originally intended as a temporary shelter. Now Pat considers this as his permanent home and plans to add on to it.

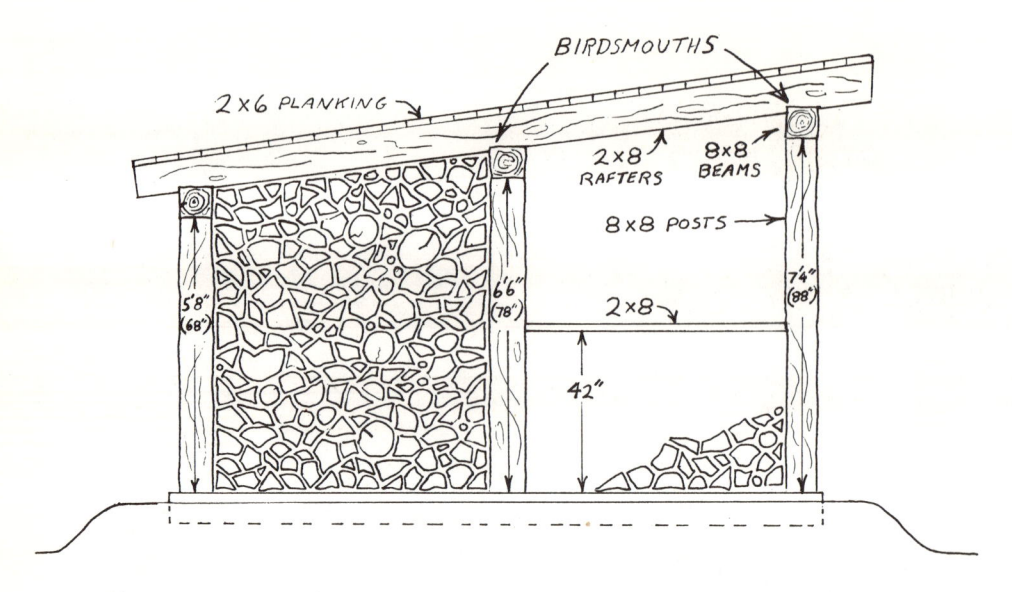

Figure 18. Post-and-beam framework for a cordwood sauna, side view.

on a floating footing or "ring beam" but it could just as easily have been built on a floating slab. (See Figures 20 and 21.)

In either case, the foundation must be poured on a pad of percolating material, such as coarse sand or gravel. Organic material and topsoil are first scraped from the site. The pad is built up about 18 inches above the undisturbed subsoil. All parts of the foundation should have at least a foot of compacted sand or gravel below. The pad thus created does not retain moisture, so the building is protected from frost heaving. At our sauna, I ran a piece of 4-inch perforated drain tile through the center of the sand pad and away to a dry well (or "soakaway") filled with stones. This drain is further protection against water build-up below the structure and doubles as a shower drain. The floating foundation is discussed in greater detail in *Cordwood Masonry Houses*.

The Temporary Earth-Shelter

Underground construction—more accurately called "earth-sheltering"—is increasing in popularity as more and more people become aware of its energy-saving and low-maintenance advantages. Unfortunately, people without building experience often assume that the technologies involved with earth-sheltering must be very complicated. They hear talk of pre-stressed concrete beams and exotic waterproofing techniques and, whoa, they're off to the woods with their *How to Build a Log Cabin* manual.

Instead, they ought to build the temporary earth-shelter, gaining familiarity with the various products and techniques involved. If the small structure (Continued on page 76.)

Figure 19. This earth-roofed cordwood sauna, with minor modifications, would make an excellent temporary shelter. The structure should last for 100 years!

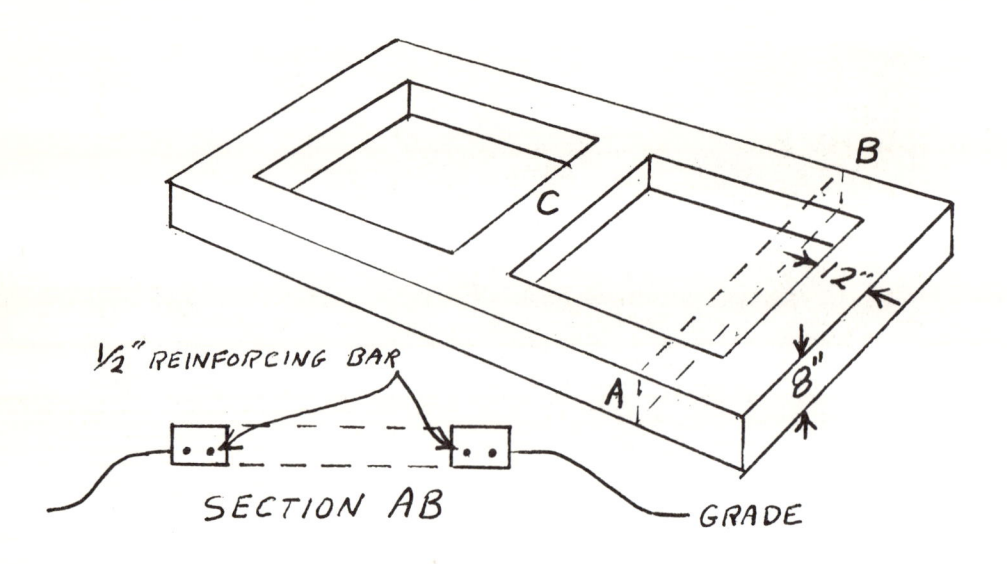

Figure 20. Floating footing or "ring beam." The grade beam "C" is necessary only for load-bearing walls. The rebar should be placed approximately three inches from the sides and bottom of the footing.

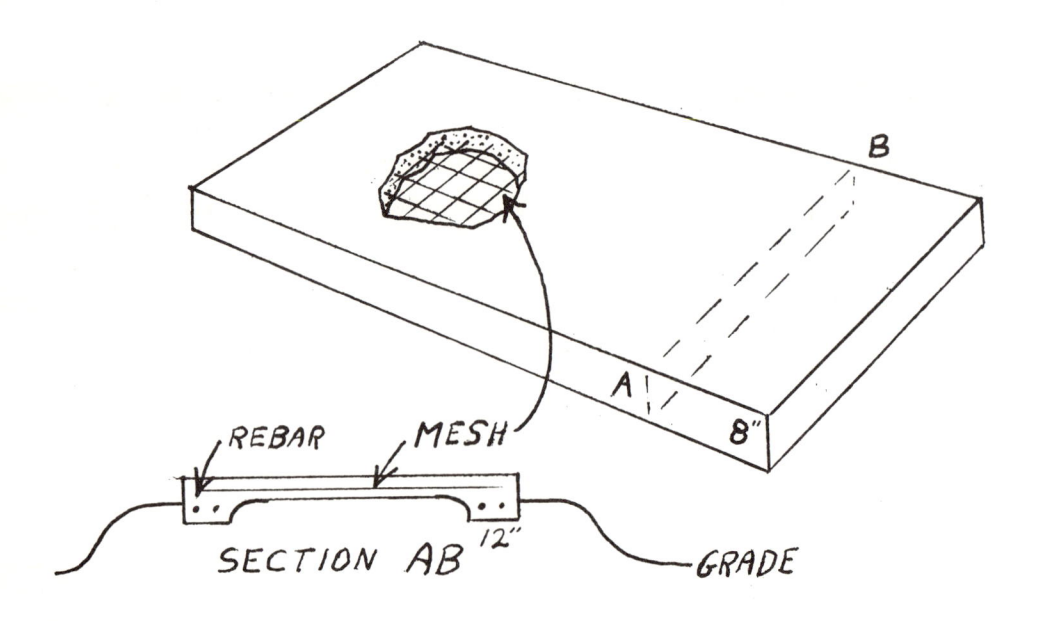

Figure 21. The floating slab can be made in one pour. The radius curve between footing and floor decreases the likelihood of shear cracking. (The floor is four inches thick.)

Figure 22. Owner/builder Steve Dorresteyn, whose log house graces the cover of this book, created these lovely yet economical effects throughout his home.

Figure 23. On this page, two views of the interior of owner/builder Steve Dorresteyn's log house. Right, note the creative use of skylights.

Figure 24. The cedar tree, plunk in the center of the room, is actually an important structural support.

Figure 25 above. This is Rich and Anne McIntosh's octagonal log cabin, built while they lived in their gambrel-roofed temporary shelter (see Figure 16). Figure 26 right. Steve Dorresteyn made appealing and practical use of a handsome piece of stained glass in his log house.

leaks, it's not the end of the world, and the builder will have a pretty good idea of what the problem is and how it can be eliminated in the house. In the pages that follow, I offer plans for a simple earth-sheltered cabin of 125 usable square feet, utilizing the same surface-bonded block wall technique and plank-and-beam roof which we employed successfully at Log End Cave.

However, I must remind you that this is not meant to be a how-to-build book. Before embarking on this project, I would advise reading *Underground Houses: How to Build a Low-Cost Home* (Sterling, 1979) for a complete discussion of the techniques involved.

In addition to all the normal uses to which the temporary shelter can be put when the main house is finished, the temporary earth-shelter—with slight modifications—could be used as a root cellar, bomb shelter, or very effective sauna. Figure 27 offers a suggested floor plan for a sauna.

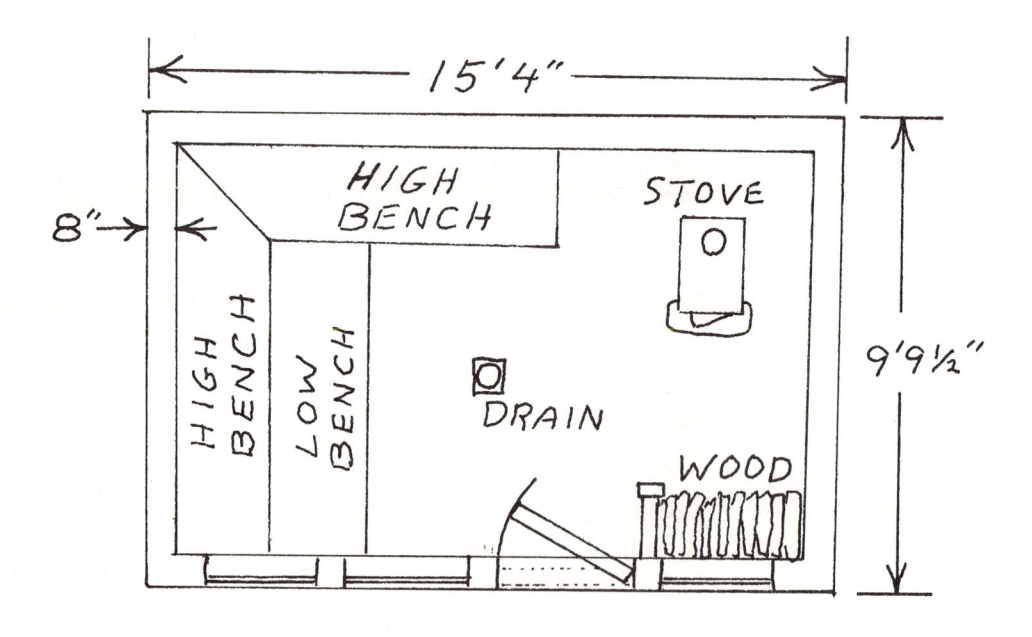

Figure 27. Above, a suggested floor plan for a sauna which would make an excellent temporary shelter that later could be converted to its permanent function.

This building would cost about $1,200 to build at this writing, $200 more if the entire structure is wrapped with an inch of Dow Styrofoam® insulation. The Styrofoam® would only be necessary if the building is to be used for human habitation during a northern winter, or if conversion to a sauna is desired. Think about this prior to construction; it is virtually impossible to retrofit an earth-sheltered structure with insulation, as its proper placement is on the *outside* of the waterproof membrane and *under* the floor.

LIVABILITY

No matter what structural form the temporary shelter is to take, *make it livable*. Invariably, owner/builders are obliged to spend a longer waiting time in their temporary shelters than they thought they would. I cannot recall a case where this has been otherwise. So, spend an extra day or two to get the place comfortable—extra shelves, a sink with gray water runoff, a good light, and so on. Organize the space carefully to maximize its efficient use.

The rough plans for the temporary earth-shelter conclude this chapter.

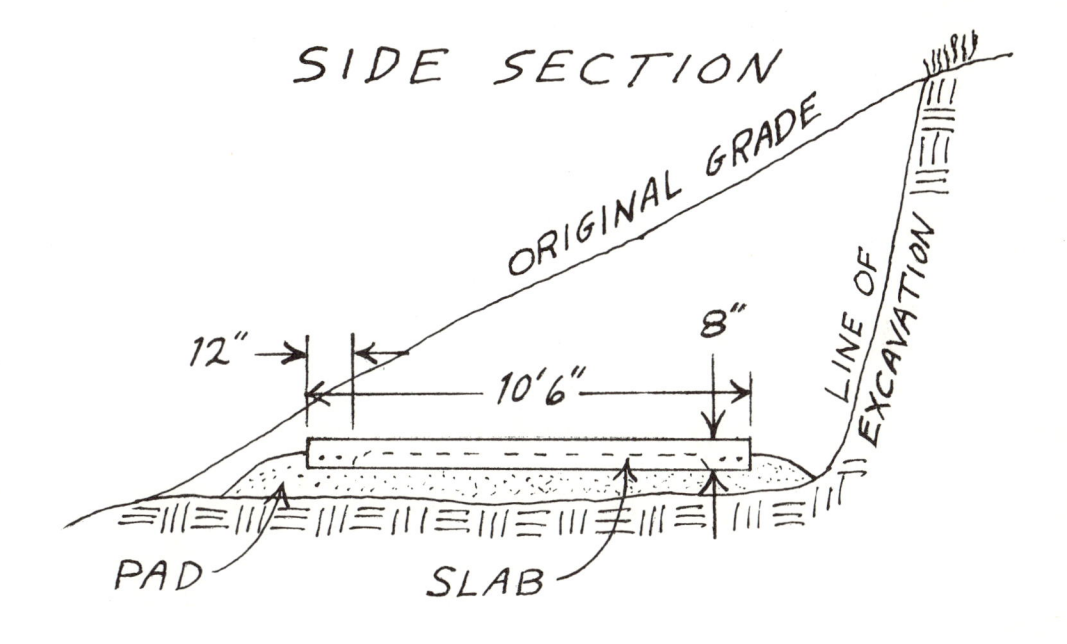

Figure 28. This drawing and those that follow should give you a clearer "picture" of the week-to-week process of construction. Above, a side section view of the first week's work—excavation, footing and floor.

FRONT VIEW

←12"

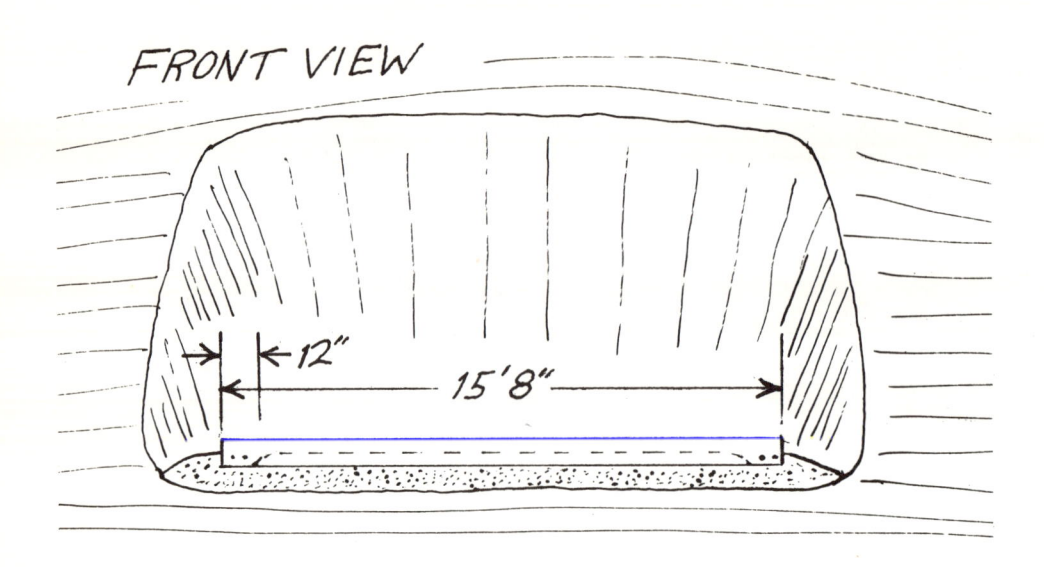

15'8"

Figure 29. A front view of the work slotted for the first week—excavation. Figure 30. Below is a top view. Remember to allow room (at least two feet) to work around the wall. The footing and floor can be poured separately or together. (See Figures 20 and 21.)

TOP VIEW

ORIGINAL GRADE

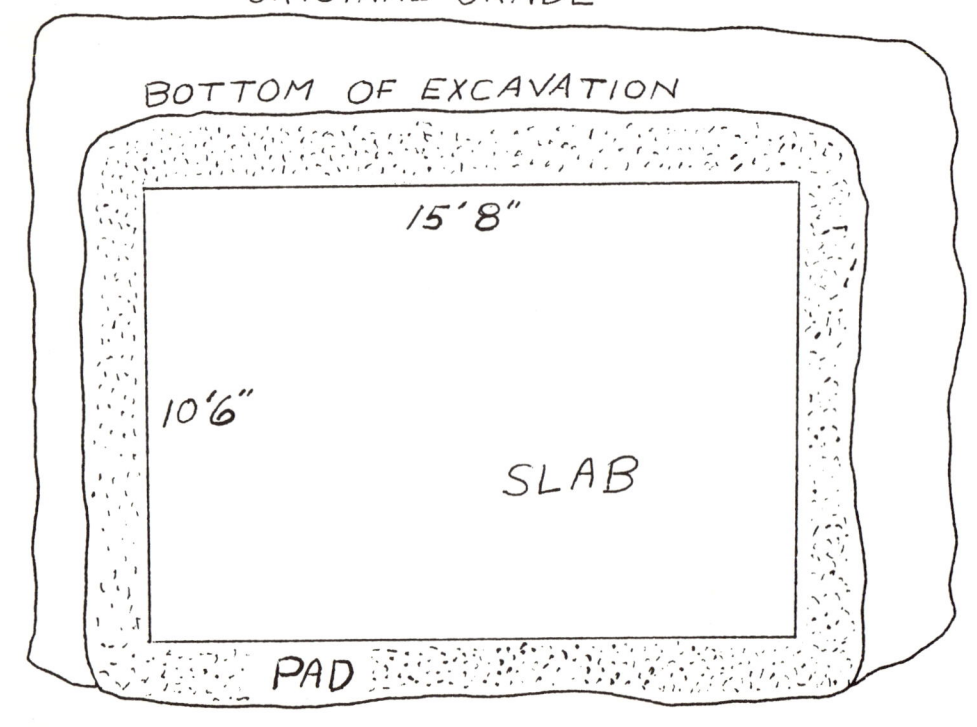

BOTTOM OF EXCAVATION

15'8"

10'6"

SLAB

PAD

PERSPECTIVE

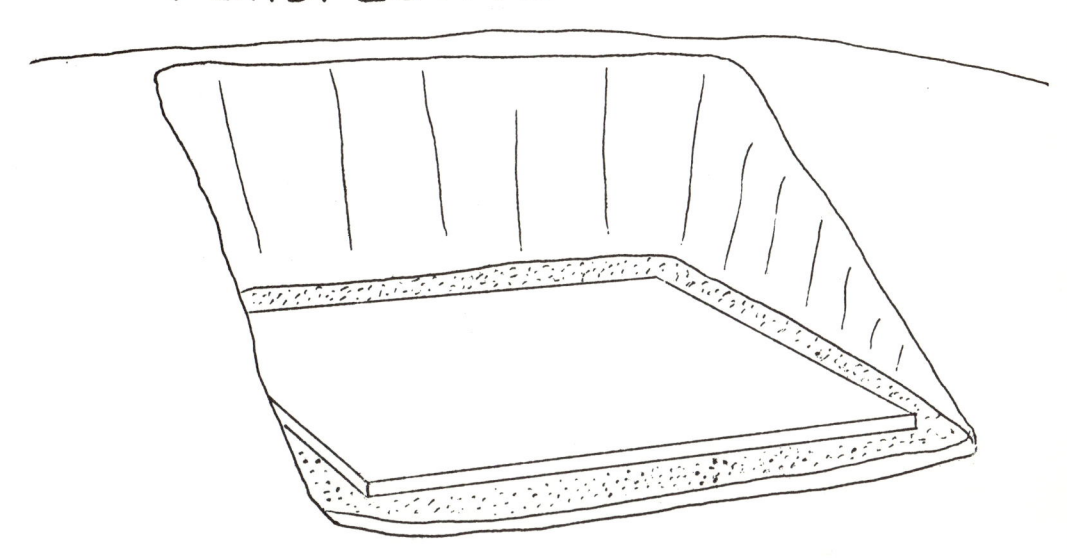

Figure 31. A tamped earth floor with a sheet of six-mil polyethylene and a carpet upon it is another option recommended by Mike Oehler, author of *The $50 And Up Underground House Book*.

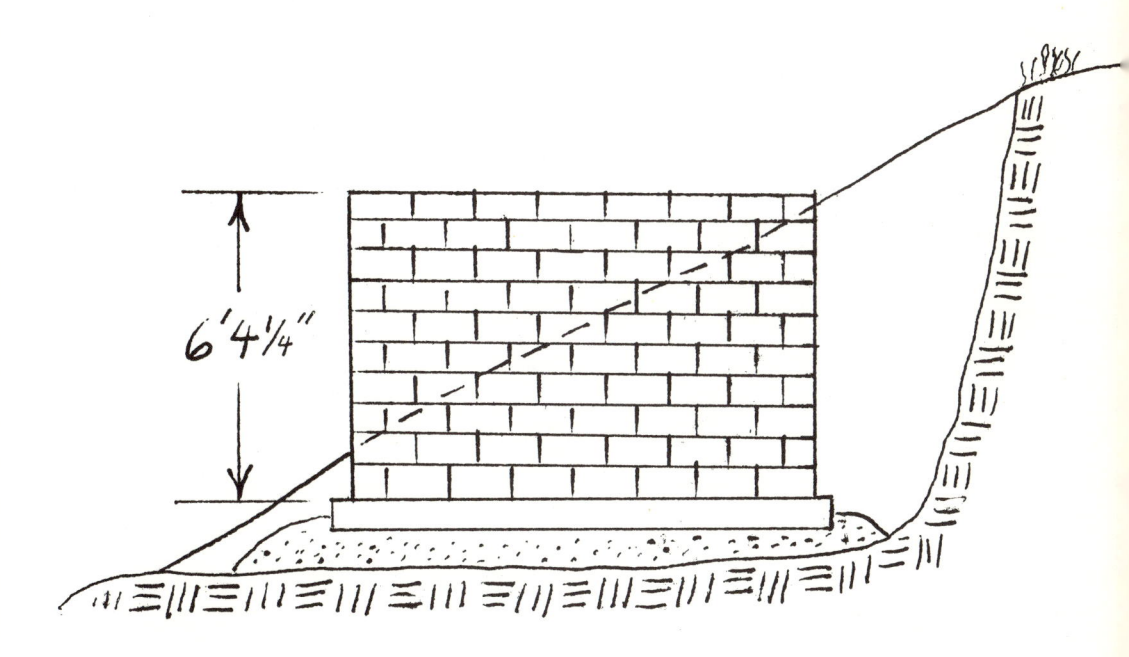

Figure 32. Second week—the walls. The east, north and west walls are stacked without mortar in preparation for surface bonding. The first course only is set in mortar to establish level. For more on surface bonding, turn page.

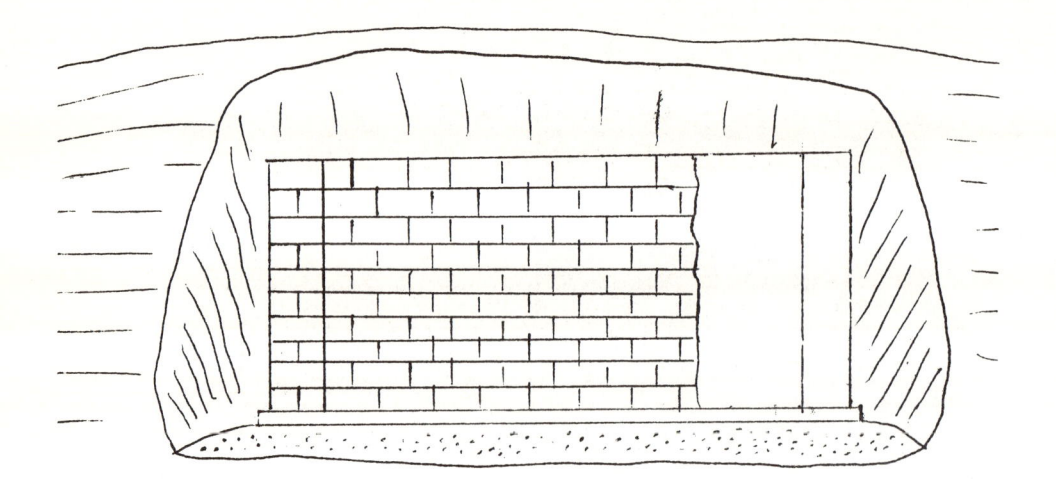

Figures 33 and 34. Surface bonding, part of your second week's work, is the application of a one-eighth-inch membrane of cement and fiberglass. Surface-bonded or "skin-stressed" concrete-block walls are stronger and tighter than conventionally laid walls. After the bonding mixture has cured, it will be strong, adhere well to the wall and will even become a waterproof coating.

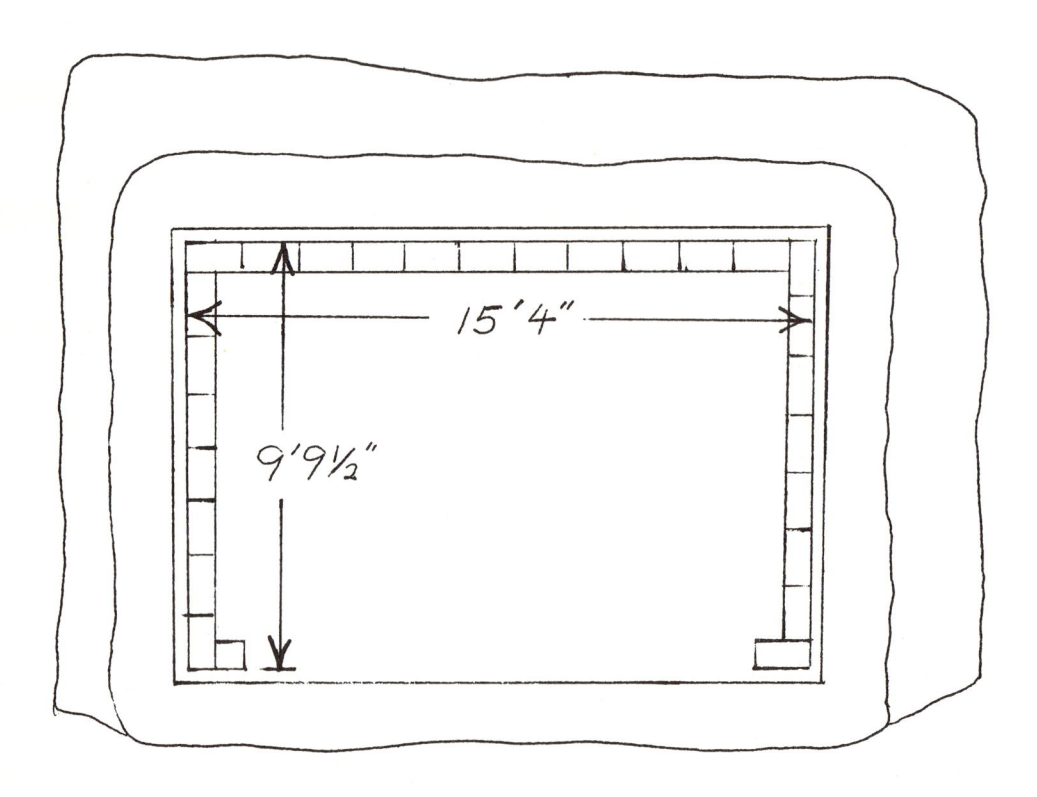

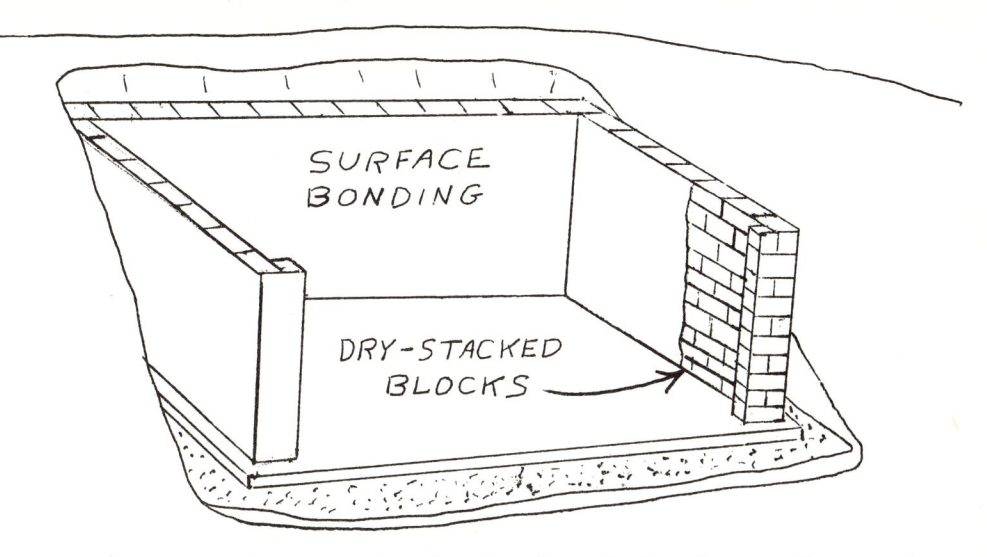

Figure 35. Above, one last view of surface bonding. The bonding-tensile strength is generally about six times that of conventionally mortared walls.

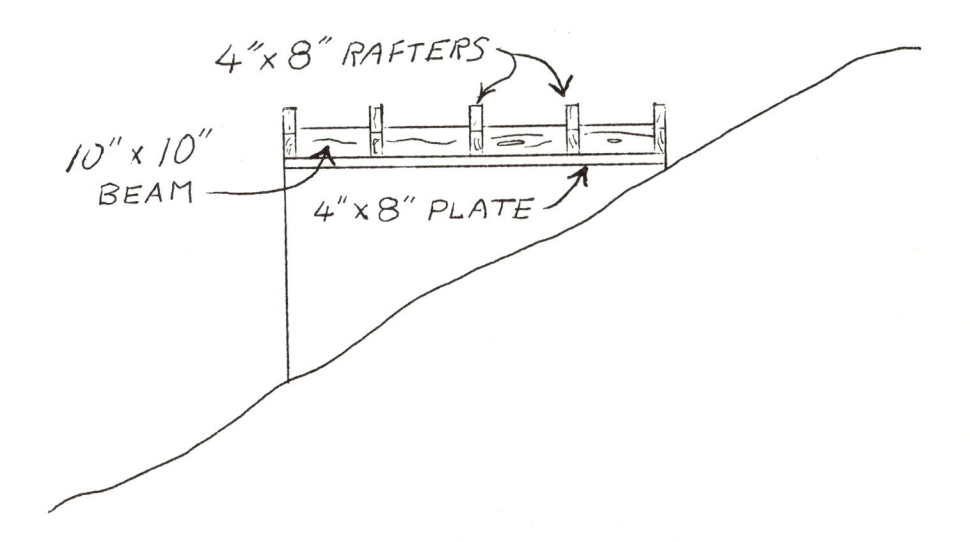

Figure 36. Third week—framing. First, the 4-by-8 wall plates are fastened to the block wall with anchor bolts. The three posts of the south wall are temporarily braced in place. For more about the third week, turn page.

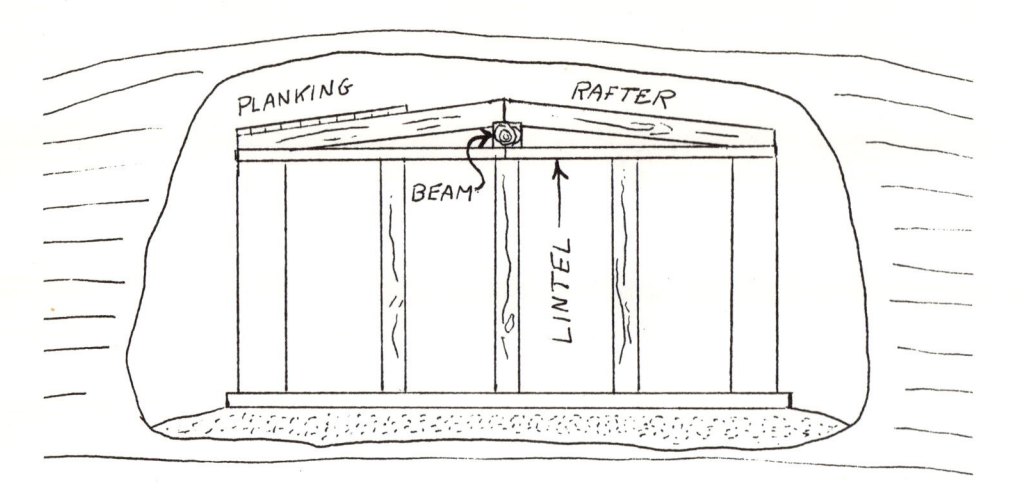

Figure 37. The three temporarily braced posts are the same height as the block wall, and can be of barn timbers, 8-by-8's, old railway ties, and so on.

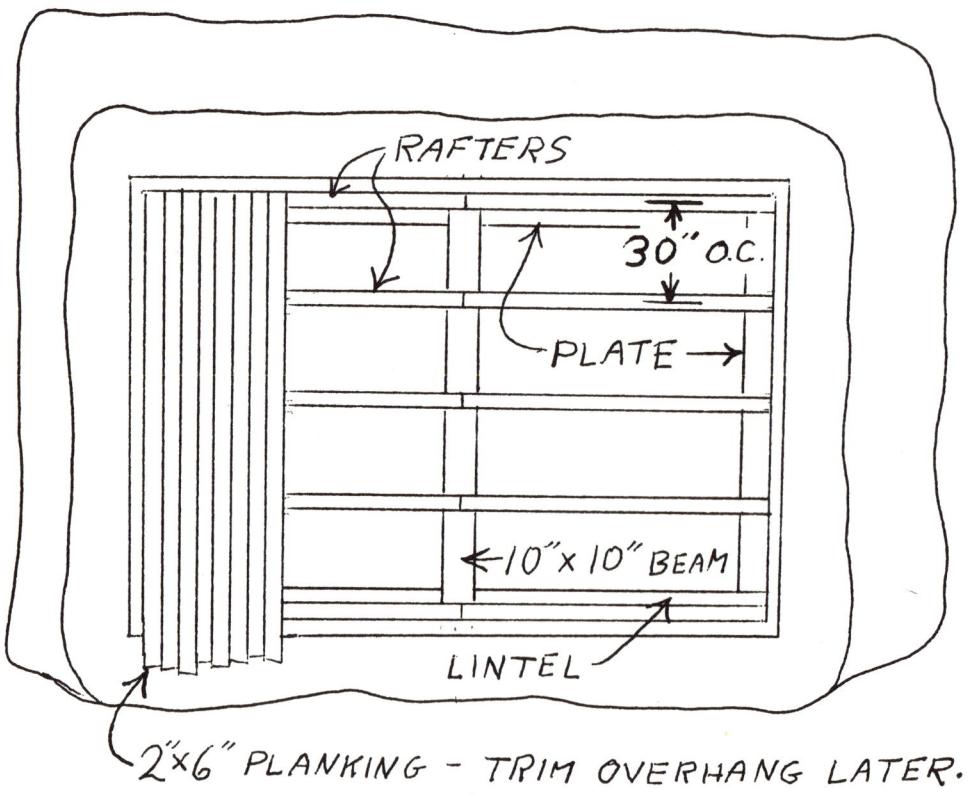

Figure 38. Now is the time to install the two 4-by-8 window and door lintels. A single 10-by-10 girder (10 feet 9 inches in length) is set to span from the center of the north wall to the center of the south wall, over the middle post. If a strong hardwood girder is used, such as oak or ash, an 8-by-8 will suffice. Otherwise, the girder can be built of four 2-by-10's.

82

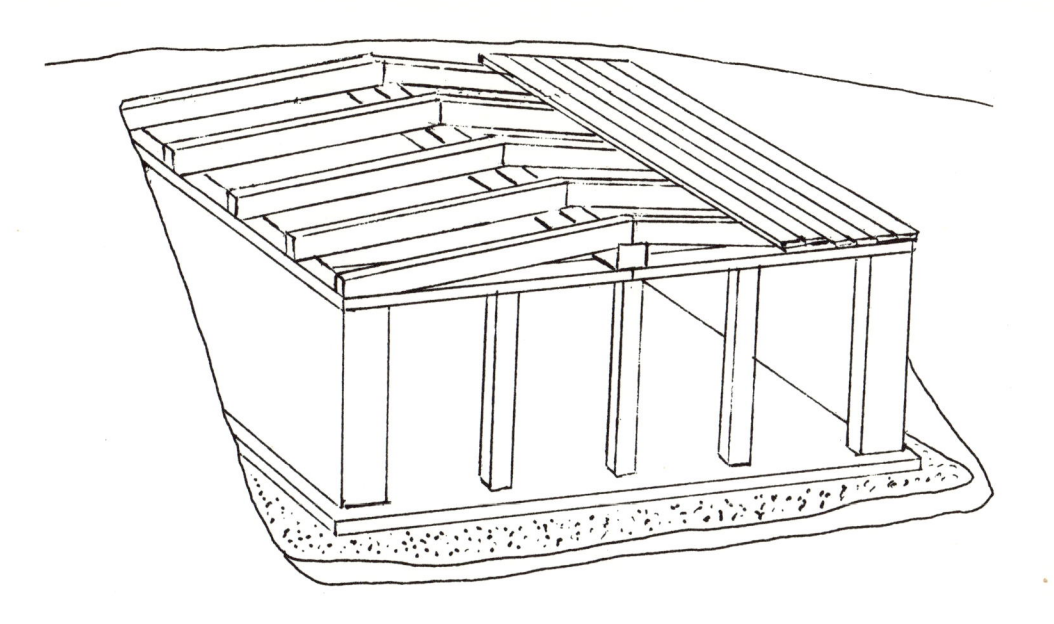

Figure 39. The ten 4-by-8 rafters—5 to each side—are set in place and then planking can commence. Planking is 2-by-6 stock, planed to uniform dimensions.

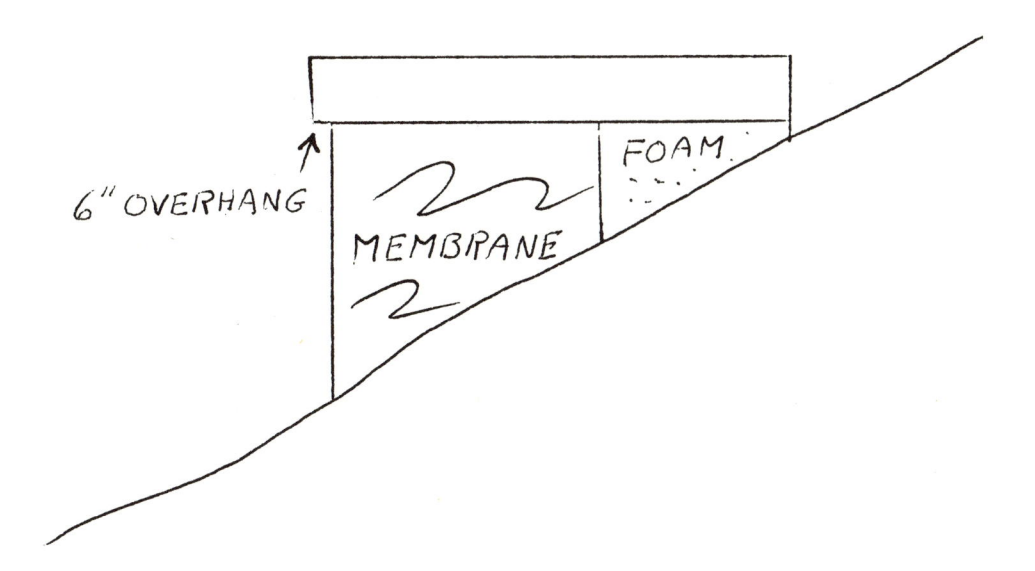

Figure 40. Fourth week—closing in. Apply the waterproof membrane and Styrofoam® (if desired) to the wall and roof. Turn page for more on fourth week's work.

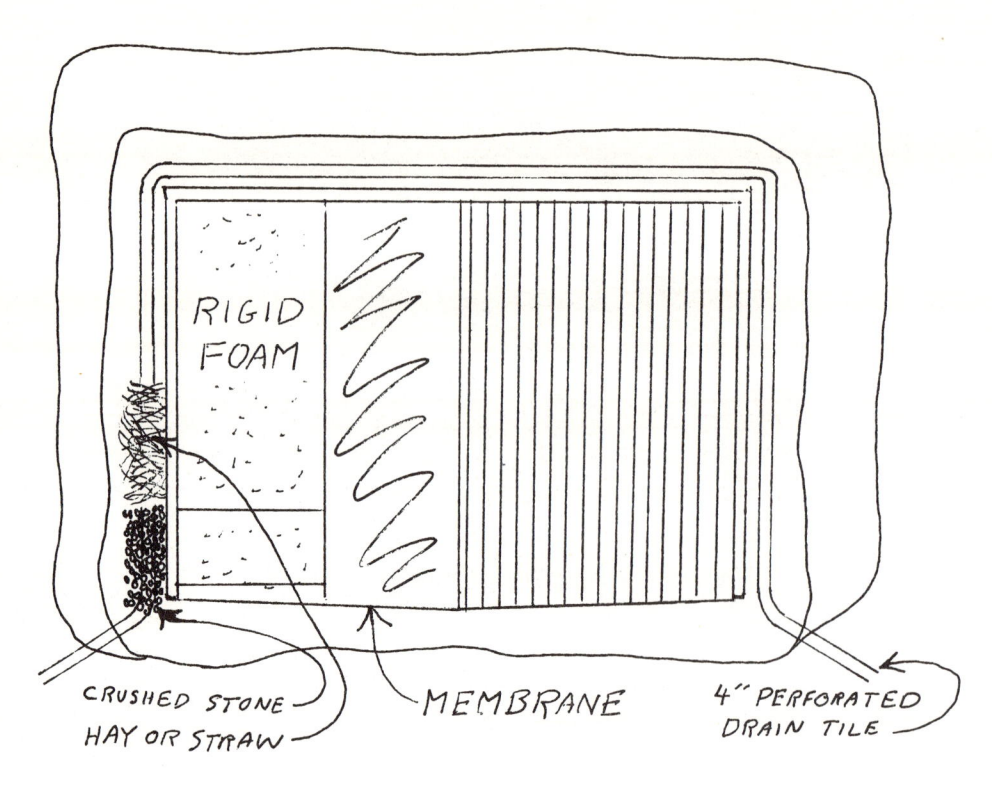

Figure 41. Now is the time to install your French drains, which consist of 4-inch perforated drain pipe set in crushed stone and covered with a hay or straw filtration layer.

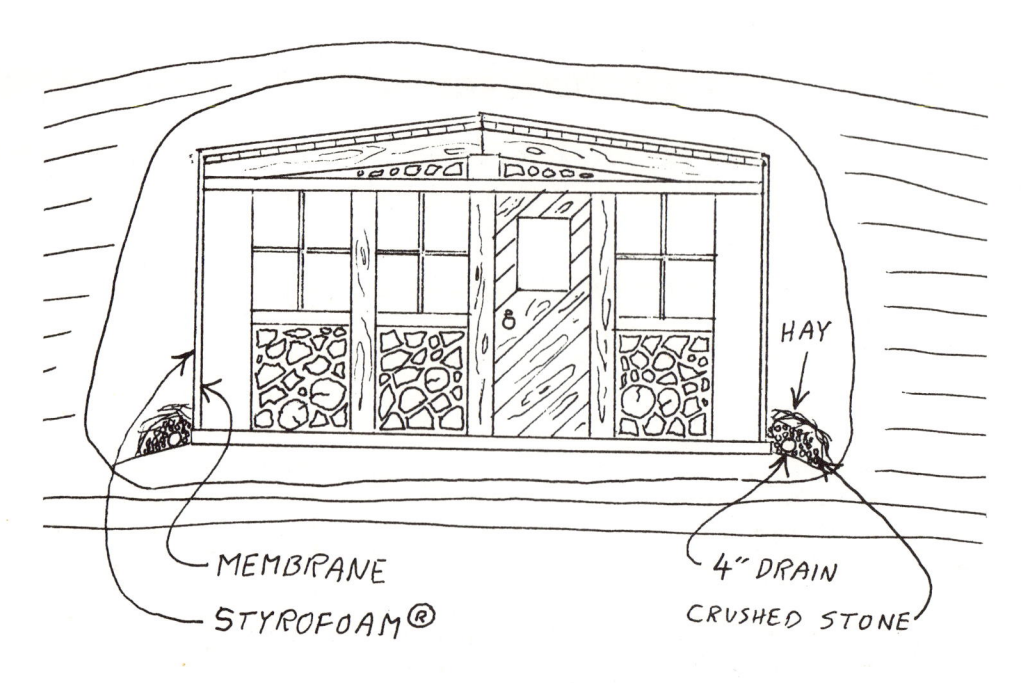

Figure 42. Install windows and doors. Almost any infilling can be used on the south wall and north gable, such as bricks, boards, adobe, stone, and so on.

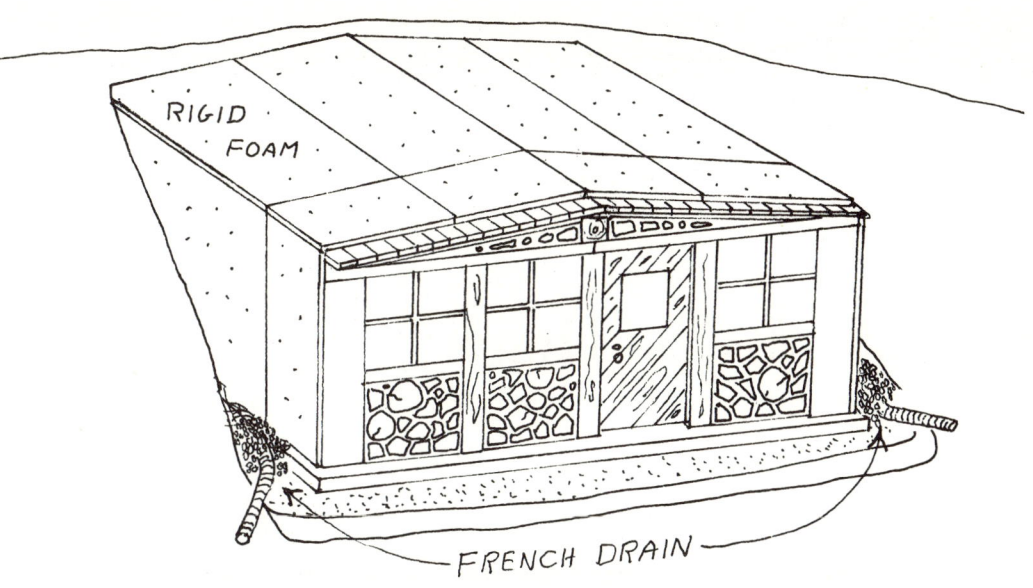

Figure 43. This illustration features cordwood masonry infilling, which is what I used for my house, Log End Cave.

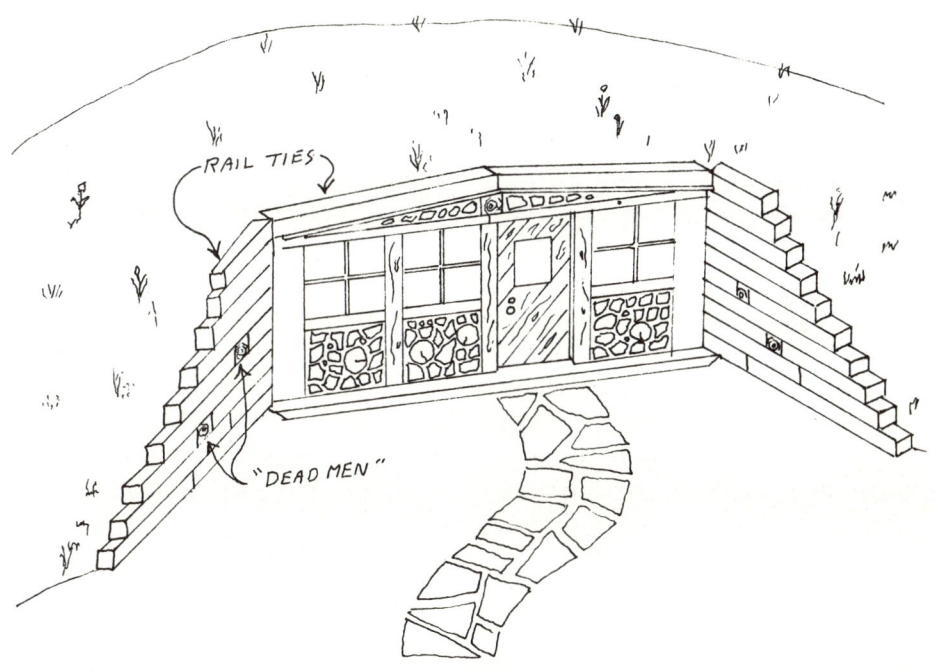

Figure 44. Fifth week—backfilling and landscaping. Old railroad ties are used here to retain earth on south side. The north "gable" is completely earth-covered. You can build retaining walls of stone, block cedar logs, landscaping timbers. The "dead men" or ties set perpendicularly into the bank help resist lateral pressure on walls. If soil has poor percolation characteristics, such as clay, a percolating backfill of coarse sand or gravel should be brought to the site. To prevent subjection to hydrostatic and frost pressures, insure good drainage behind retaining walls as well as behind walls of building. Supports a 9-inch non-snow earth load.

5

The Low-Cost Home

Building your own low-cost home involves the integration of many strategies, but, just as a house requires a strong foundation, building plans require the strong foundation of advance research. If you already have your land and know what kind of house you want to build, your job is very much easier. Using the subject index at a good library, you can find out what has been written on the required building techniques for your design choice. It is important that a reference work on building be clear, precise and well illustrated. Good illustrations are often worth more than the accompanying text.

Figure 10 is an excellent example. Adapted from Bob Easton's simple drawing, this illustration is really good enough to build from. With a quick glance at the original, we were able to fashion our own scheme for our temporary shelter. As a technical writer, I would require several thousand words to explain the construction of this simple shed, but thanks to Old Tom, I was able to show it quickly.

Research should not be limited to reading. Observing, helping and talking with experienced builders are all valuable methods of learning. In fact, book knowledge alone is rarely sufficient preparation, as books cannot adequately impart the human side of construction.

Fine, but what if you're new to an area, and don't know any builders? Drive up and down country roads and keep your eyes open for signs of building activity. Very likely, there are others in your area starting out on the same adventure. Or perhaps you know of or notice a local farmer who's adding a shed or barn extension. Don't be shy. You're their new neighbor, and here's a way for the acquaintance to be made. Usually, folks will be glad to accept a helping hand, especially if you are pursuing a common goal. You're doing each other a good turn. It is true that some farmers are too independent to allow a stranger to help, but they might not object to your stopping by every few days to take note of the progress and construction techniques. And they might allow you to study the construction of some of the older buildings on the farm.

Another possibility for gaining experience is to take a job as a laborer on a construction crew.

Seminars, workshops, conferences and building schools are all excellent sources of useful knowledge. They cost money, but they are almost invariably worth more than the investment. The flow of information and ideas is very intense. (See Appendix for a list of several good schools and seminar programs.)

What Kind of House?

People ask me, "What's the best kind of house to build?" and I'm sure that they expect me to say either cordwood masonry or underground, because of my special interests in these fields. They really want to see *which* of the two I will name. My reply is always this: There's no universal answer to your question. The right construction for me is not necessarily the right for you. There are excellent houses of stone, timber framing, logs—even earth. Use indigenous materials. Satisfy your own senses of comfort and aesthetics. Unless you've got lots of ability or lots of money, or both, keep it small and simple.

Often this answer is unsatisfactory. "No, no, I mean, which is better, cordwood *or* underground?" they persist.

I won't let them pin me down. I tell them that cordwood masonry is the cheapest and easiest way of building a house in *this* area, but it's not without its problems—finding dry cordwood, for example. As for underground, I wouldn't live in any other but an earth-sheltered home. This does not restrict the builder to any particular construction technique, however. Poured walls, stone, surface-bonded block, even cordwood masonry are all viable underground modes. Roof structure can be wood or concrete. Roof surface can be earth or something more conventional. (Our next house will be a round cordwood structure with earth-bermed walls and a sod roof.)

Planning

In their enthusiasm, people often begin planning their house—usually floor plans—before doing the research outlined above. And they almost always plan their houses before they've even bought their land. I advise against both of these approaches, although my advice on this probably will be ignored. We were guilty of both procedures ourselves six years ago, but Jaki and I have learned since.

Prior to actual house design, you must have an idea of the house style which you intend to build: framed, log, stone or whatever. This choice most often will be a personal value judgment based on the aesthetic, practical and economic balance of the appropriate styles for the area. Prior research into the availability of materials, cost and energy relevance can avoid deadends and disappointments later on. This is one of the reasons why the site is so important to planning.

You might be a young couple with a romantic vision of a little log cabin in the woods, for example, and you plan accordingly. Then you find that the only wooded land which you can afford has softwoods averaging only 6 inches in diameter. The two possibilities here are: a) building a cabin with insufficient wall thickness to keep you warm, or b) you end up buying a pre-cut kit or logs from another lot to satisfy your romantic vision.

In the former case, the energy cost may prove to be unacceptable, even though the romance might keep you warm for a while. In the latter case, the dollar cost of the home takes a big jump and some of the economic advantage is lost, particularly that potential which derives from the use of one's own indigenous materials.

Sometimes the strong desire for a particular type of home can enter into the land search, by narrowing the pertinent properties quite severely. Although this limits the search somewhat, the results can be well worth it if good use of the property's natural resources is made. Lovers of stone houses can really do well in this instance, as stony land is usually priced considerably lower than nonstony land.

Another example of too early planning that I often encounter occurs in underground housing. Every kind of house has its limitations, and the primary limitation of earth-roofed houses is that large, internal spaces are extremely expensive due to the structural cost of supporting tremendous roof loads on long spans.* People whose limited knowledge of earth-sheltering is that it contributes to energy-efficiency, will come to me with plans including an 18-by-24-foot living room, and the intention of placing 3 feet of earth on the structure: I tell them, "Well, if you cut your earth load down to 6 or 8 inches, the minimum necessary to maintain the green cover, *and* put a stout post in the center, you could have a room of this size at reasonable cost."

"Oh, but we can't possibly have a *post* in the middle of the room! And we *know* it can be engineered, because we just saw it in *Better Homes and Gardens*."

"Oh, the calculations are not that difficult," I say. "But the elimination of that single post and the use of 30 inches of superfluous earth on the roof will double—and possibly triple—the cost of your home."

Long faces usually result from this explanation. They could be avoided if

---

* AUTHOR'S NOTE: Earth-sheltered housing does not *have* to have an earth roof, as I am well aware, but the advantages are quite compelling, and include aesthetics, longevity, ease of summer cooling, superior runoff control, better harmony with the ecology, sound and radiation protection and some degree of extra insulation, especially by the retention of the snow cover, worth about R1 per inch.[24]

the building style is researched and the building plans tailored to the structure and not the structure to the floor plans.

A similar kind of mistake is made by people who design their house, then find their land, *then* set out to tailor the site to the house plans—a violent course of action—instead of adopting the more gentle and harmonious strategy of shaping the house plans to the site. Western architects and builders can learn from their oriental counterparts here, and also from the examples of folk architecture down through the ages.

Okay, assuming that the proper investigation into the chosen building style has been made and the building lot is owned and in harmony with the chosen building style, beneficial planning can commence. There are four rules which the first-time owner/builder should keep in mind throughout the planning stage.

1. Keep it small.
2. Keep it simple.
3. Tailor the floor plans to the structural considerations and the available materials, not the other way around.
4. Consider all the living systems at the design stage, not just the shelter requirement.

Let's look at these four rules in depth.

Keep It Small

The cost of building materials and the dollar cost of energy have already had an impact on house size. After years of increase, the average new house size began to drop slightly at the end of the last decade, at about the same time that gas consumption began to decline. The average new home is still in the 1,700 square foot range, however, and banks are reluctant to loan money on "small homes" (less than 1,500 square feet), fearing a low resale value.*

---

* AUTHOR'S NOTE: When people start harping on "resale value," I can't help relating the true story of my friend, Russ, who found himself on a beach in Morocco without shelter. He gathered together some sticks, stones and banana fronds, set up a tentlike framework of the sticks, roofed the structure with the fronds, and covered the whole thing with a sheet of plastic, for which he'd paid $2. The stones kept the plastic from blowing away, and Russ bermed the little house with sand (to the extent which the framework would allow) in order to take advantage of the "coolth" below the hot beach surface.

After living in comfort in his $2 earth shelter for a month, Russ decided that it was time to move on. Wishing to leave the environment in the same state as he'd found it, he decided to dismantle the structure. A fellow beach-dweller, who did not have a place of his own, said, "Don't knock it down. I'll buy it from you." Russ sold the shelter for $6, a 200 percent appreciation in a month. Not wanting to take advantage of the buyer, Russ threw in a frying pan for good measure!

This does not affect the mortgage-free home, but it reflects the kinds of obstacles put in front of people who wish to build and plan appropriately to current energy and economic realities.

Over 300 years ago, Thomas Fuller said, "Better one's house be too little one day than too big all the year after."[25] This is true again today, after an unfortunate period of wasteful use of the planetary capital.

But building small just for the sake of building small serves no purpose, either. A family's space requirements fluctuate. Young couples with a low budget can live comfortably in a small house which would not be suited to a family with three teenaged children. A small house can be expanded, as need dictates and personal economy allows.

Although economics is the obvious reason for building small—the cost of a house is pretty much proportional to the square footage—it is not necessarily the most important one. The important reason for building small *is to get the thing completed!* Inexperienced builders, even those with plenty of money, should not tackle a house larger than 1,000 square feet. There is a very real danger that the place will never be completed. Owner/builder Donald Pellman, in a perceptive article for *Country Journal,* puts it this way:

> How long does it take to complete a house? It is difficult to say on the basis of the homes I visited, because none of them were finished, although the oldest had been standing for four years. All the builders agreed that this represented one of their biggest changes in attitude. Without experience, no one can comprehend how long it will take to put a house together, they say. Surprisingly, these people estimated their costs with fair accuracy: budget overruns were typically only twenty-five percent, and the average house (about 24' by 32', with one-and-a-half or two stories) cost between $600 and $10,000 (in 1976). But almost everyone admitted that the *time* overrun was at least one hundred percent.[26]

I have seen couples break up over incompleted houses, and overambitious projects is one of the major causes of incompleted houses. There are lots of reasons why people think they need to have a big house, aside from bank propaganda and outmoded zoning regulations. Two of the more prominent I call the *overreaction syndrome* and *bedroom mania.*

THE OVERREACTION SYNDROME

Jack and Jill have been cooped up in their little apartment or house trailer for so long that all they can think is, "When we build our house, there's gonna be plenty of *space!*" They've got lots of time to plan; paper and pencils are cheap. They finally get started on their 3,000-square-foot masterpiece. The possibilities from there, in descending order of probability, are: (1) There is a great enthusiasm to begin with. After about six months, money, energy and patience run low, then run out. Jack and Jill split up. (2) After a while, Jack and Jill perceive that they've really bitten off too much. They move into one-third of the place. "Someday we'll finish the rest," they say. (3) They pull it off, as planned. I have heard rumor of this, but have yet to witness it through personal contact.

BEDROOM MANIA

The functions of a bedroom are to supply a peaceful venue for horizontal resting of the body and as a storage area, generally for clothes. The bedrooms in most American homes could be divided in two and each would still serve the purpose. Sure, lots of other considerations come into the planning: building codes again; an adjustment, perhaps, of the individual value system; planning a small bedroom to accommodate furniture. One thing is certain: the larger the bedroom (or *house,* for that matter), the more unnecessary "stuff" one accumulates.

As for number of bedrooms, this is largely a function of individual thinking. Americans seem particularly concerned with the issue of privacy. Every kid has got to have his bedroom and then we throw in one extra for the pot: the ubiquitous "guest room." In the reality of most family situations, the guest rooms are used less than 10 percent of the time. The living room of our house has a convertible sofa, so the room functions nicely as a guest room, too. A side benefit of this strategy is that guests are less likely to overstay their welcome than if conditions are made just too comfortable.

I realize that the above views are highly personal, but it is my intent to show that there are many trade-offs when it comes to building a mortgage-free home. As children get older, and the need for privacy becomes more genuine, an addition can be built to meet the changing circumstances, and it is likely that the family will be better able to afford this extra space a few years down the road.

THE ADD-ON HOUSE STRATEGY

One of the popular and successful strategies open to you is that of building a small, affordable core—typically in the 500- to 700-square-foot range—and to build affordable additions as they become truly necessary. The Brasacchio and Light case histories in the next chapter are good examples of this strategy successfully employed.

Sometimes the temporary shelter will serve as the core for the completed house, either by plan, or by evolution, as in the case of the Light residence. However, get one part of the house completely finished before moving on to the next part. Living in a house under construction puts tremendous strains on a marriage. If you can retreat to a clean and uncluttered living area, this "refuge" may prove invaluable on all living fronts.

There are two schools of thought with regard to add-on houses. One is to have some specific expansion plan in mind at the initial design stage. The other says to let the house grow organically as such needs arise. My observation is that both plans will work, and you should therefore tailor your strategy to your personality. If you have an organized, analytical mind you may be happier knowing that you're working toward some specific end or goal, while a more spontaneous individual might feel cramped by such a plan, preferring free creative rein throughout. My own approach is a kind of hybrid between these two ideas. I allow myself to be locked into certain structural requirements, largely because I lean toward the use of massive timber framing and cordwood masonry, but spontaneous creativity is given its outlet through the use of special design features within the structural framework. Rather than feeling restricted, the use of stone and wood masonry gives me a sense of freedom—to create special textures, alcoves and designs. Creative carpenters like Pat Duniho and Ron Light (see Chapter 6) express themselves in a similar way by their use of log and framing techniques.

Although I have not used the add-on strategy myself, I offer one caution with regard to its use, and it may be that my not using the strategy is a direct result of not having been so cautioned myself. Log End Cottage and Log End Cave would be quite difficult, structurally and aesthetically, to add on to. Both structures were designed to be complete in themselves, and, therefore, no thought was given to expansion potential. When Log End Cottage became too small, thanks to the addition of just six pounds of little boy, our choices were to attempt a difficult addition or to build a new house.

Well, our decision to build a new house was not *entirely* predicated upon this situation; it really would have been somewhat easier and cheaper to build an addition to the Cottage. But our desire to cut our heating requirements from seven cords of wood down to three, and my new interest in earth-sheltering, was enough to set our course of action. Not always one to learn from my first mistakes, I went and designed Log End Cave also without expansion potential. In fact, the Cave is even more locked into its present size and shape than the Cottage! While it is true that the Cave's 910 square feet of useful living area is quite sufficient for a young family of three or four, it is a little disappointing to know that this size is pretty much what we're stuck with. Small is beautiful, yes, but have a care for the future.

If you're planning an earth-shelter, be aware that, unless expansion is specifically addressed at the design stage, underground houses are very difficult to add on to.

A few years ago, a young couple with several small children visited us to discuss earth-sheltered housing. They insisted that they absolutely *had* to have 1,600 square feet of living area, *minimum*. The trouble was that they could only afford 800 square feet. I advised them—I always give better advice than I take—to build the 800 square feet that they could afford, leaving the east wall of concrete block insulated, but not backfilled. As they were both making good money as truck drivers, they could afford to complete the other 800-square-foot module two or three years down the road. This could be accomplished by reusing the rigid foam insulation—as long as it was protected from the sun's ultraviolet rays—and utilizing the internal masonry wall as a thermal flywheel and effective noise buffer between one side of the house and the other. (See Figure 45.) The result: an energy-efficient debt-free home. The trade-off: Two or three years of less than the desired living space.

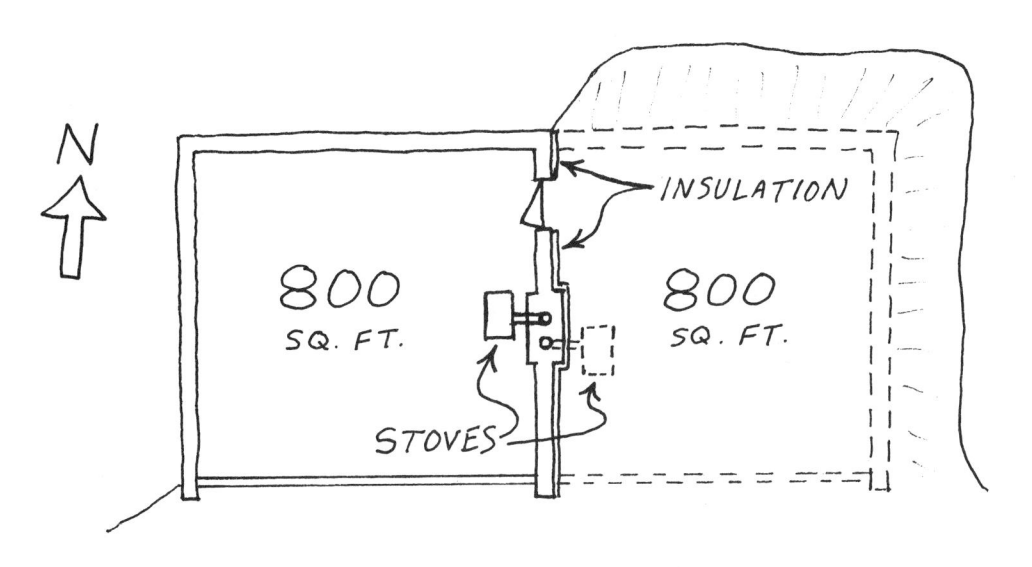

Figure 45.

Keep It Simple

This second rule of sound and economical construction should not be confused with the first, "Keep it Small." A small house can be hopelessly complex, and a large house can be wonderfully simple. The next few pages deal with some of the considerations pertinent to simplicity of construction.

KEEP TO ONE STYLE

There is a style which suits your personality and pocketbook better than others; once you have found it, stick with it. If two house shapes are to be intersected in some way, let there be some unifying force to the architecture— a constancy in choice of roof or wall materials, for example. A hodge-podge house always looks like a hodge-podge house.

AVOID DIFFICULT LINES

If you think they are tough to draw, wait until you try to build 'em! Keeping simple lines is of particular importance if you're inexperienced. Gambrel, hip and valley roofs (and dormer additions) should be avoided on the roof line, for example. (See Figure 46.) Sunken living rooms, complex stairways and split levels can all cause grief in the interior.

Geodesic domes, polygonal houses and yurts may each have a strong appeal, but these projects should be undertaken with the understanding that (1) the finish work is long and tedious, (2) furniture is designed on the premise that

GAMBREL　　HIP

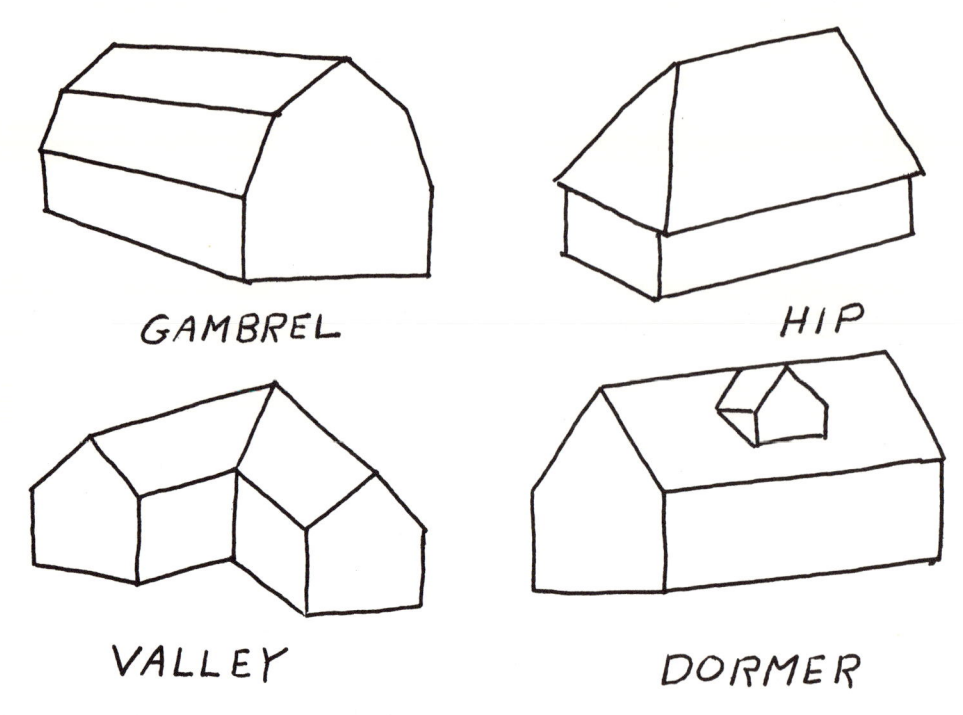

VALLEY　　DORMER

Figure 46. Some examples of complex structures to be avoided.

gravity runs perpendicular to the horizon, and (3) it's unlikely that there will be local people experienced in these techniques to offer advice when you get in trouble. As Donald K. Pellman says, "You can't ask a local carpenter how to put a sod roof on a yurt."[27] (A yurt, by the way, is a circular, latticework domed dwelling with slanted roof beams, skylight, unique portability and adaptability.) If in doubt, build a model of the intended structure. If you can't build the model, don't tackle the house.

THE ROUND HOUSE

A cylindrical house of masonry, by the way, is *not* difficult to build. At first, this might seem contradictory to the previous comments on domes and yurts, but it is not. The walls of domes and yurts are not perpendicular to the horizon, whereas the walls of a cylinder are, aiding construction (including window and door installation) and providing better use of space in the upwards direction. (See Figure 47.) So-called "primitive" builders, such as ancient European stone masons, the Hopi Indians of the American Southwest and present-day African tribes, all chose cylindrical buildings for the ease of their construction and their functionality. That the buildings are also beautiful is no coincidence, if, indeed, form follows function.

The efficient use of materials in cylindrical houses is demonstrated in Figure 48. It is an indisputable function of plane geometry that a circle will enclose 27.3 percent more area than the most efficient rectilinear shape (a square) of the same perimeter. When compared to the more common rec-

DOME YURT CYLINDER

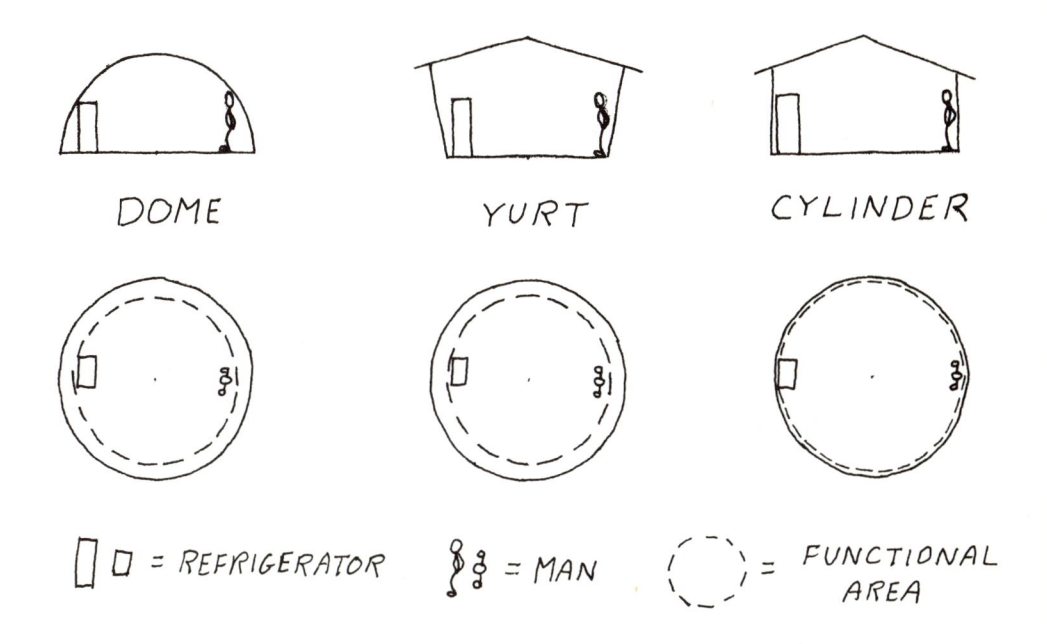

☐ ▫ = REFRIGERATOR   ⸺ = MAN   (⸺) = FUNCTIONAL AREA

Figure 47. A cylindrical house makes better use of floor space than does a dome or a yurt, as you can see from the drawing above.

tilinear shapes that Western builders use for their shelter, the gain of round building is more like 40 percent.

Another advantage is that a round house is easier to heat than a square or rectilinear house with the same floor area. The economy of wall area is part of the reason, of course, but the lesser wind resistance offered by a curved wall is also important. And if a radiant heat source, such as a wood stove, is placed at the center of the structure, there will be no cold corners, as all points on the circumference will be equidistant from the heat source.

Building a round masonry house is easier than building a rectilinear structure of the same materials. (See Figure 49.) First, a pipe is set plumb at the center of a round floating slab. An assembly consisting of a bull's ring, a nonstretch line or wire and a plumb bob or heavy spike establishes uniform distance from the pipe. If the stones, bricks, blocks or log-ends are always set to the extended plumb bob assembly, the wall will be plumb, round and smooth—without constant use of a level. As the walls go up, it is necessary to raise the bull's ring on the pipe correspondingly or the wall will begin to slope inward like a dome.

The most difficult part of round house construction is the roof, but this is only slightly more awkward than a rectilinear roof, if it is planned with a view to efficient use of materials. The radial rafter system shown in Figure 50 is probably the easiest where a roof pitch is required, but builders of adobe structures in areas of low rainfall successfully use almost flat roofs and parallel rafters called "vigas." (See Figure 51.)

A: 38.2' INTERNAL AREA: 1146 SQ FT

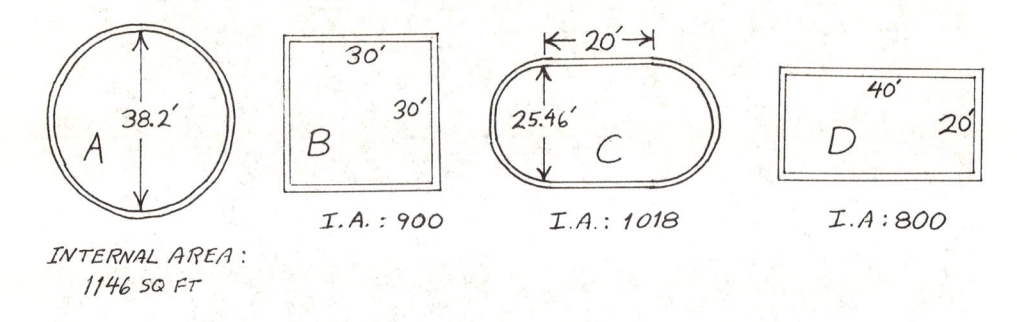

B: 30' × 30' I.A.: 900

C: 20' × 25.46' I.A.: 1018

D: 40' × 20' I.A: 800

Figure 48. The perimeter of each of these houses is the same: 120 feet. However, the curved-wall shapes are a much more efficient use of masonry.

Cordwood Masonry Houses (Sterling, 1980) goes into detail on round house construction and the commentary is equally applicable for stone, block, brick or adobe houses as for cordwood. Figure 52, originally appearing in my previous book, shows a 1,000-square-foot round cordwood house with an earth roof. This style is inexpensive and easy to build where wood is the indigenous material. The earth roof is optional and requires much heavier framing than for other roofs.

AVOID BASEMENTS

It is surprising how many people in the North continue to view a basement as a prerequisite to house construction, despite the fact that in a low-cost owner-built home, a basement will eat up about half the building budget or half the man-hours or some combination of these, while providing low-quality space which will be used less than 10 percent of the time.

We made the same mistake at Log End Cottage. The basement took a full third of the total cost of the house, and the labor was considerable. Its only useful purpose is to store garden produce, which could have fit into a small

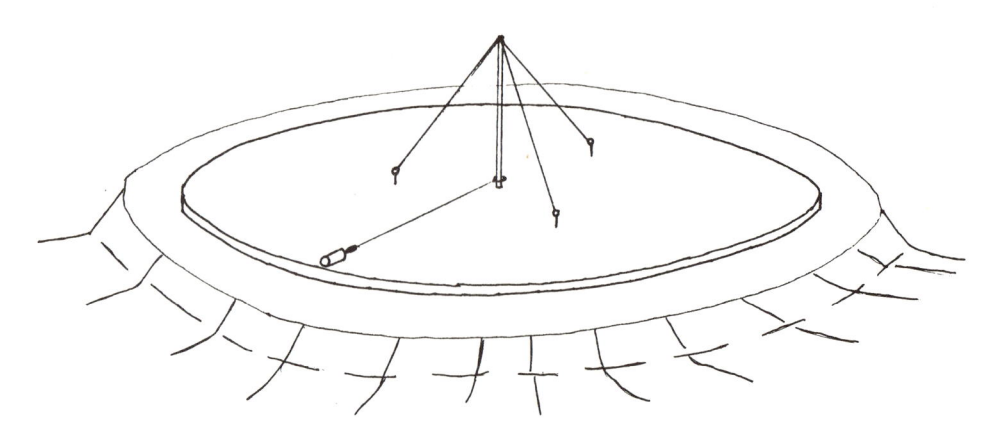

Figure 49. With the center pipe plumb, laying up masonry to the line assembly is a breeze.

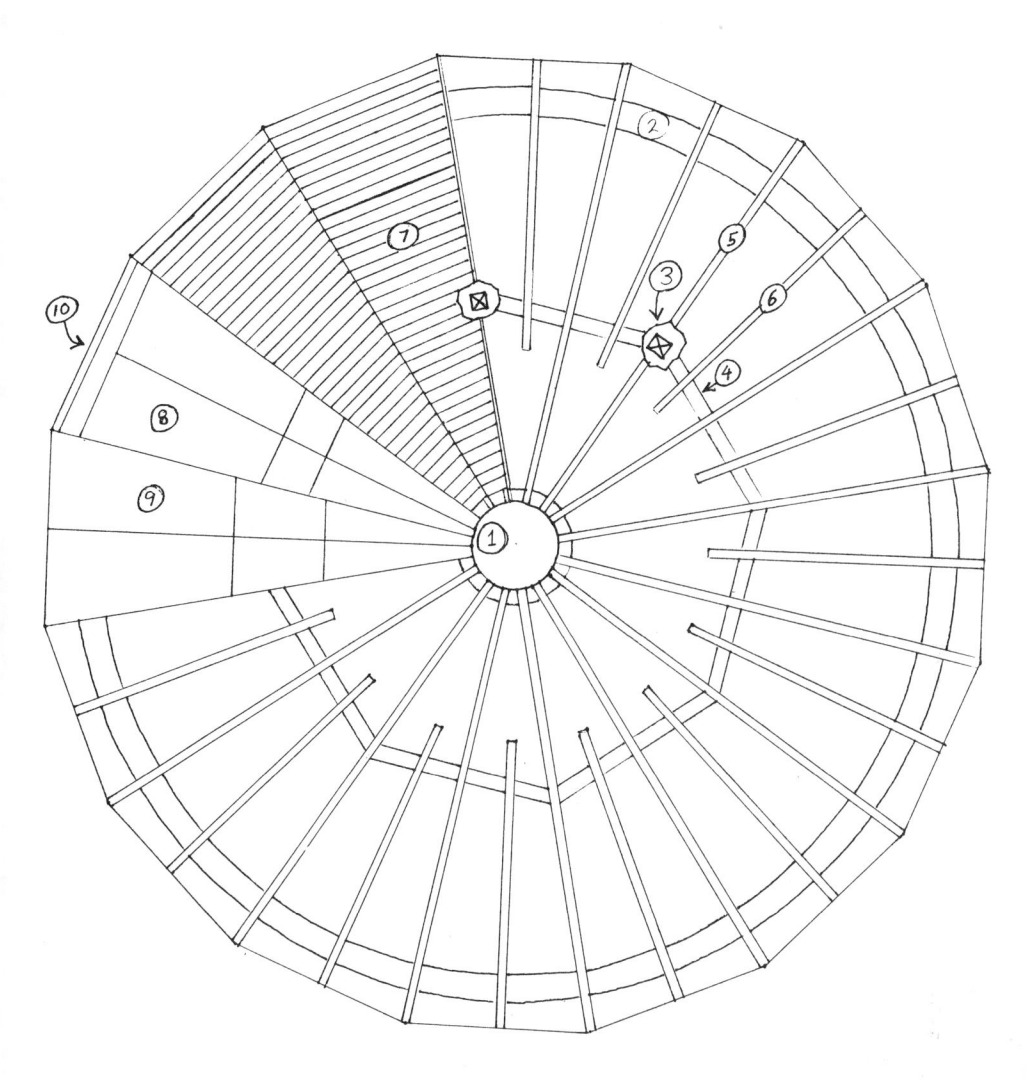

Figure 50. Structural plan of the Round House. (1) Stone heat sink; (2) 16-inch cordwood wall; (3) Post locations, eight in all; (4) Eight-by-eight beam, part of octagonal support structure; (5) Primary rafter; (6) Secondary rafter; (7) Two-by-six planking; (8) Rigid foam insulation; (9) Hardboard; (10) Two-by-four or four-by-four, to match depth of insulation.

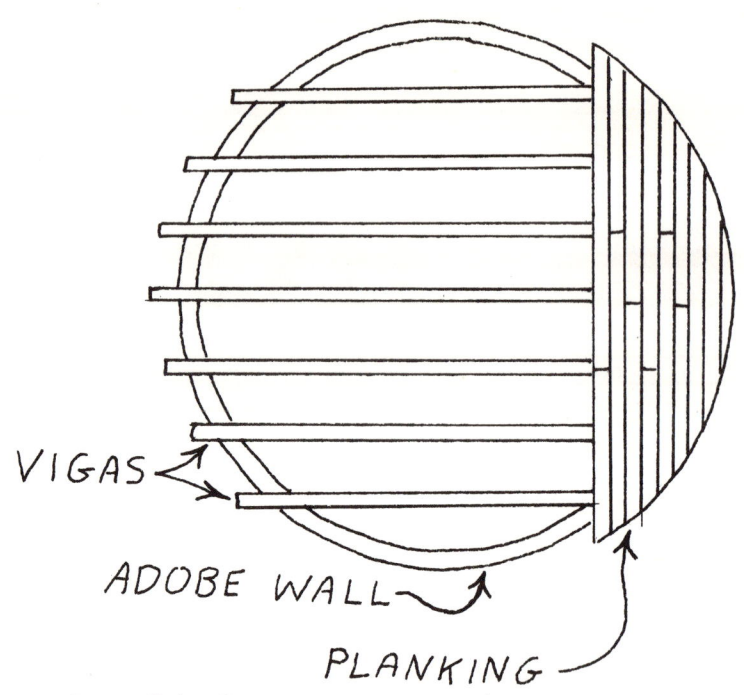

Figure 51. A parallel rafter system on a round structure, common with adobe construction.

root cellar under a floating slab and connected to the pantry by a trap door, at a combined cost (slab *and* root cellar) of less than half that of a full basement. Most functions, *including heating,* are best enclosed in the house proper, not in a basement. Pure and simple, basements are not cost economic, require familiarity with an additional structural system (which is why most owner/builders contract their basement out, usually at a high cost relative to the house) and provide low quality habitat for almost anything except mushroom propagation.

You may be wondering how an earth-sheltering advocate can be so critical of basements. After all, what's the difference? Fellow undergrounder Mike Oehler says:

> An underground house has no more in common with a basement than a penthouse apartment has in common with a hot, dark, dusty attic.
>
> A basement is not designed for human habitat. It is a place to put the furnace and store junk. It is constructed to reach below the frost line so that the frost heaves don't crumple the fragile conventional structure above. It is a place where workmen can walk around checking for termites under the flooring, where they may work on pipes and wiring. Its design, function and often even the material from which it is built is different from an underground house. A basement is usually a dark, damp, dirty place and even when it is not, even when it is a recreation room, say, it is usually an airless place with few windows, artificially lighted and having an artificial feel.
>
> An underground house is not this at all.[28]

In his next chapter, Mr. Oehler tells what an underground house *is*. Our own experience bears out his commentary. The basement at Log End Cottage is dark, damp, dingy and of little practical value. Log End Cave is light, bright, dry, cheerful and practical.

If you still feel that a basement is a requirement in your owner-built home, I would advise you to read Rex Roberts's commentary on this matter in his excellent book, *Your Engineered House* (M. Evans and Co., 1964), pages 20 to 23. Then, and only then, build a basement.

THE CAPPED BASEMENT

Many owner/builders in our area build a "capped basement" and move into it as a temporary shelter until they can afford to build a house on top. The primary advantages of this strategy are that these structures, if properly built, are easy to heat and carry a low property tax valuation. The main disadvantages are that they are often short of light, are humid, leaky, and singularly unattractive to look upon. Proper glazing, waterproofing, and insulating techniques will add considerably to the cost of the structure, but the result will be a much more valuable and useful space.

Appropriate Floor Plans

The third planning rule: Tailor the floor plans to the structural considerations and the available materials, not the other way around. I can best illustrate this rule by example. I have already told of structural problems which can be encountered in earth-sheltered houses where the structural plan is made subsidiary to the floor plan, but the error of this line of thinking is not limited to earth-shelters. Economy of construction is largely predicated upon economy of building materials. We have said that a circular structure makes the most economic use of masonry materials. This might not be true with regard to standard frame construction, utilizing studs, plywood, sheathing, and the like. A square house may be more practical in this case. Long narrow houses, projections and courtyards are all inefficient uses of material if it is the builder's desire to keep the square-foot cost down.

Look at Figure 53. Both houses have exterior dimensions of 24 feet by 36 feet, a common and quite sensible size chosen by first-time owner/builders. House A has 864 square feet on the ground floor. The designer of House B took the same basic dimensions, but decided that he wanted a floor plan that puts the master bedroom and the children's bedroom in separate "wings," with a courtyard in front of the house. Moreover, he likes the look of the structure better than House A. Fine, these are all valid value judgments and, after all, everybody's different. But you should know what these value judgments will cost, in terms of time, money, ease of construction and energy efficiency.

House B, with its 12-foot by 12-foot courtyard, loses 144 square feet of interior living space and adds 24 feet to the house perimeter. Each perimeter foot of wall encloses 5 square feet of interior floor area, whereas in House A,

Figure 52. The Round House.

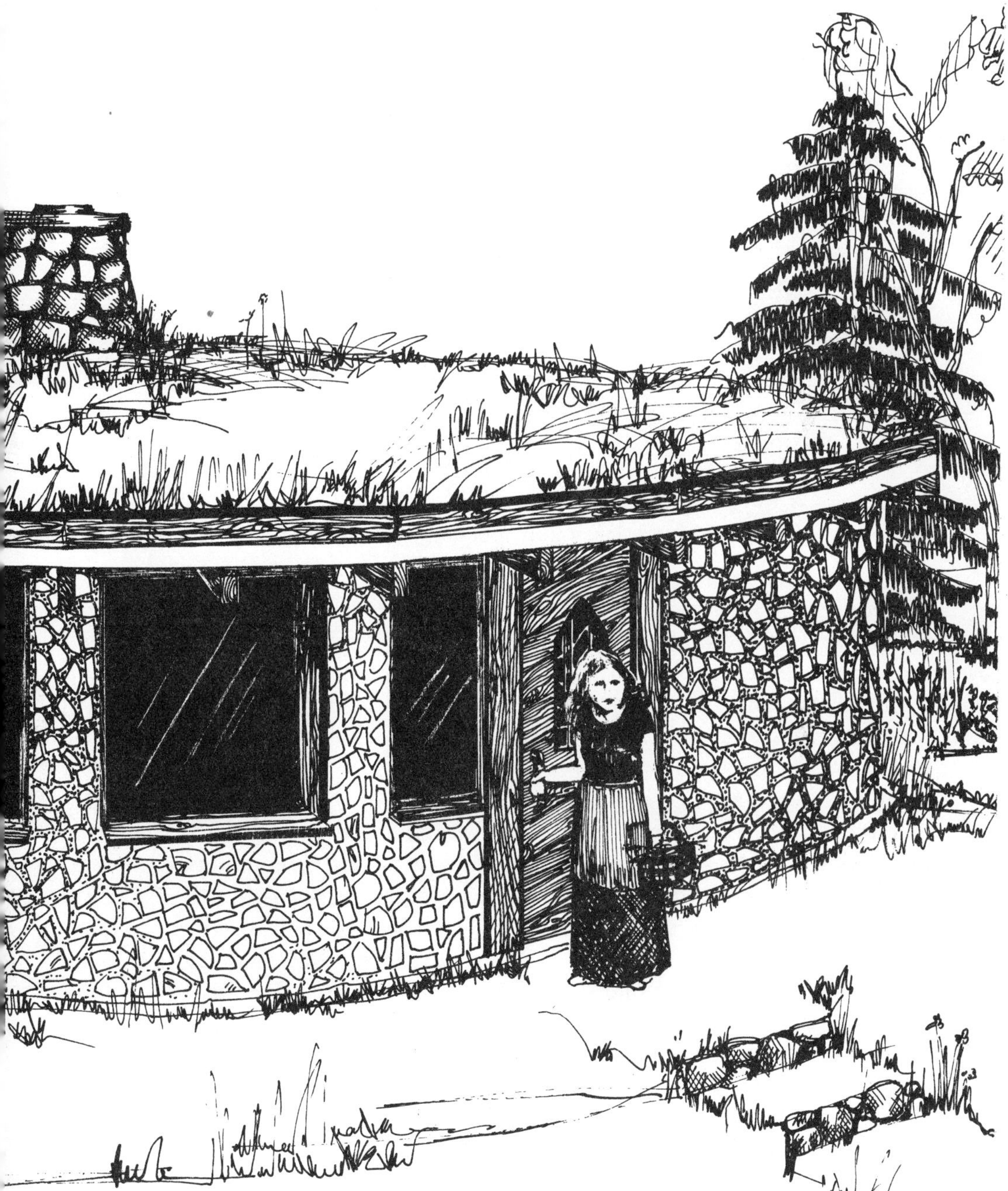

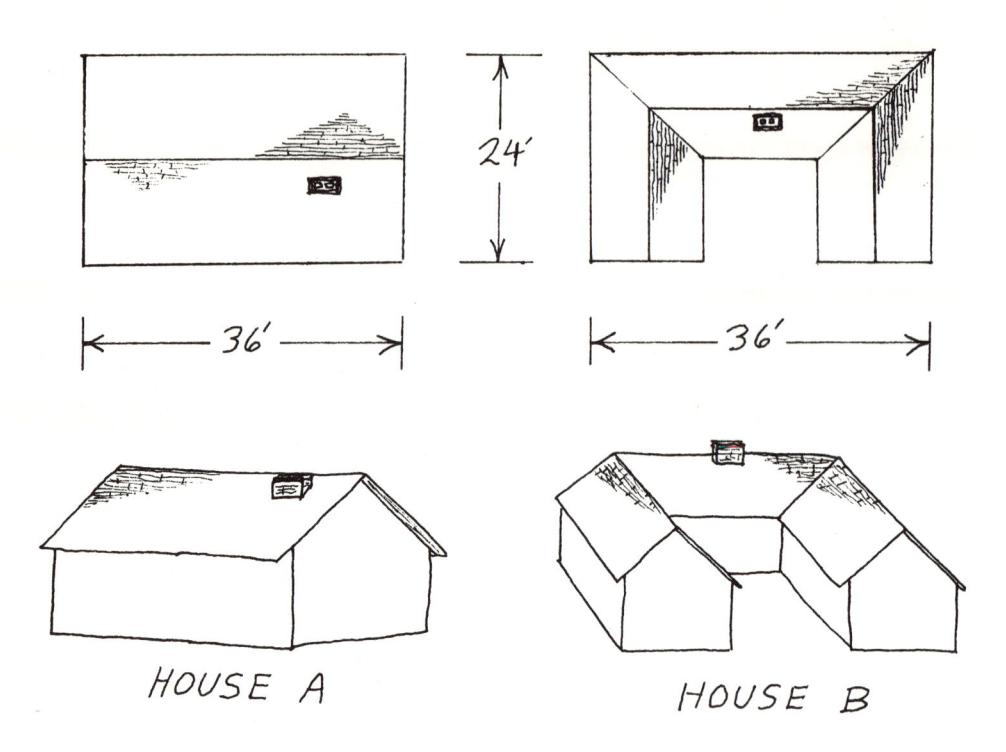

Figure 53. Design considerations must be weighed against cost and energy factors.

7.2 square feet of floor area is enclosed by each perimeter foot. House B is only about 70 percent as cost efficient in terms of external walls. The energy saving characteristics of the respective walls have a similar relationship, too, because heat loss is proportional to skin area, other factors being equal.

There's more. Roof B, though slightly smaller than Roof A, is much more difficult to build and has more wastage of materials due to the varying sizes of rafters and the unusual shapes of the decking and roofing material. The foundation is the least inefficient part of House B's construction, but even here the features desired in the house must be paid for with other trade-offs. If a concrete floating slab is poured, there will be a very slight decrease in the amount of concrete used in Slab B—don't forget the extra perimeter, where the footings are 8 inches thick—but this minor gain is more than offset by more difficult, time-consuming and expensive forming. And Slab B will not be quite as strong in terms of frost and settling resistance because of its shape.

Finally, House A is much more conducive to using upstairs space than House B. Sleeping lofts and attic storage space would both be more useful. If 2 to 4 feet is added to the wall height of House A, considerable expansion potential exists. A family might live on one floor for a year, finishing up bedrooms on the second floor as time and money allow. The same extra wall height in House B will not yield anywhere near the same gain because of the size and shape of the attic space.

If, after weighing up the extra time, dollar cost, energy cost and the building nuisance factor of Design B, you decide that you can indeed build the house, that you can afford to build it and that the desired features justify the cost, then you should carry on and build it. You have made an educated decision, you'll probably succeed in the endeavor and you'll have built the house you truly want.

I've deliberately stacked the deck against House B to illustrate certain design considerations. I do not mean that everyone should run out and build house A. A design like House B is often the result of fitting a structure to a desired floor plan or location of spaces. Perhaps the same goals could have been achieved in another way, by fitting the floor plan to an economical and easy-to-build structural plan. Figure 53a shows a way that the practical goals of Design B could have been achieved at a great saving of time and materials. A south-facing solarium replaces the courtyard, rendering it more useful throughout the year. The house still has separate bedroom areas, but the main frame and roof structure is greatly simplified. This plan may not satisfy your yearning for a ranch-style house, but it might make the pill a little easier to swallow.

Floor plans should always take into consideration the availability of materials. If standard framing is to be employed, wall and floor dimensions must take standard building material dimensions into consideration.

Plywood is 4 feet by 8 feet, for example. If plywood is to be used in the construction, on either walls or floors, a house which is 25 feet by 35 feet with a wall height of 9 feet is not a good plan. A design which is 24 feet by 36 feet by 8 feet fits in extremely well with the use of plywood.

Dimensional timbers come in 2-foot increments. You can't buy a 13-foot floor joist off the store shelf, so you would have to buy a 14-footer and waste a foot. (And sometimes 14's are hard to find, too.) It is generally more cost-

Figure 53a. An economical and easy-to-build structure.

effective to keep to 12 feet, if possible. Beyond 12 feet, the larger dimensions, such as 2-by-10's, become increasingly expensive *per foot* because larger trees are needed to yield the extra length. Only the small end of a forest log is of interest to the sawyer. The greater the length and taper, the more the waste.

RECYCLED MATERIALS

This seems the appropriate time to discuss the use of salvaged building materials. When you are able to score a real bargain on some major structural component, don't throw the opportunity away because, say, the salvaged joists are only 11 feet long and you want a 24-foot-wide building. If there's enough, or *nearly* enough, to do the job, change the plans to a 22-foot width. At Log End Cave, I virtually designed the basic house shell around three old 30-foot 10 inches-by-10 barn beams that I bought for $90 the lot. Buying equivalent beams new would have cost several times as much and they wouldn't have had the same character.

Another example: a local manufacturer of insulated glass panels—thermopane—is sometimes left with units which have been made the wrong size for a job or were not collected by the customer. The manufacturer's options are to dismantle the units and use the glass again on smaller orders, or to sell the units at a discount of 60 percent or more. We used double-pane windows throughout Log End Cottage and Log End Cave. The total cost for all 11 windows at the Cottage, including one made to order, was only $161 in 1975.

The greenhouse attached to our shed is constructed of rough-cut 2-by-4's, silo planking for the roof, and 20 units of 1-inch Thermopane (¼-inch plate glass, ½-inch air space, ¼-inch glass), each measuring 15½ inches by 42 inches. Every one was cut ½ inch undersize for another job. I bought the lot for $96, less than 20 percent of their value, and a couple of other useful units were thrown in for nothing. These window units will be recycled yet again at Earthwood, the house we are building at this writing.

If you're stuck into a rigid plan, you won't be able to take advantage of this kind of bargain, even though a savings of $1,000 or more might be in the offing. That's why the strategy is to *design around the available materials,* be they indigenous to the site, salvaged from an old building or obtained at an auction or a going-out-of-business sale. (Pat Duniho's house, built for $1,000 complete, is an excellent example of this strategy. See Chapter Six.)

PLAN WITH THE SITE

Most people plan their houses assuming that the site will be flat. To be sure, flat sites are the easiest to build on. In some cases, however, the best site on a lot—best in terms of the visual impact, solar and wind orientation or access—might not be flat. The two options are to change the plans or to change the site. It is easier—and kinder to the planet—to change the plans. By the same token, split-level homes, which might make perfect sense on sites with a particular slope, could be grossly out of place on a flat or very steep site.

On a steep building site, two options are: 1) build a two-storey earth-sheltered home or 2) build a house on pillars.

Get professional engineering advice when building any kind of house on a steep slope. There are tremendous pressures that work unrelentingly to make everything flat. A builder's job is not to beat these forces, just to hold them off for some desired period of time. Even the Great Pyramids, *even Mount Everest,* will eventually succumb to gravity and erosion. But let's not go looking for trouble with shoddy structural engineering. If in doubt, don't.

I meet lots of people who think that an earth-sheltered house must be built on a south-facing slope. Well, that's a nice situation, to be sure, but it is not the only siting option. By shallow excavation and berming, an earth-sheltered house can be built on a flat site.* However, do avoid low-lying sites with drainage problems, such as bogs and river flood plains. Also, avoid north-facing slopes, unless heat dissipation is desired, such as in the extreme South. Undergrounder Stu Campbell says, "The perfect exposure for a window meant to collect solar radiation is 15 degrees west of true South, but 20 degrees to either side of this point is still excellent."[29]

The more beautiful the site, the simpler the structure should be. When it comes to aesthetics, we will not improve on Mother Nature at her best. The house should not take the eye away from the setting, but harmonize with it. This is why it is so difficult to design a house before the site is chosen.

AVOID FANCY FINISHING

This means paint, spackling, wallpaper, trim, floor finishing, bathroom tiles, carpeting. These can add from 25 percent to 75 percent to the house cost at little, if any, practical gain. Design them out of the plan as much as possible. Structural materials should be chosen which have built into them the desired surface qualities. Painting and staining wood serves no purpose outside of changing the color, and in most cases the change does not constitute an improvement. Stain always darkens wood. Paint must be repainted and repainted and repainted. Wood left to its own devices, *even on a floor,* will start out pleasing, stay pleasing, and get more pleasing with age. Above-grade wood does not need to be "preserved." People pay fancy prices for weathered barn boards, but hesitate in putting new rough-cut lumber on their walls. Weathered barns were built of new materials, and have been beautiful *all* their lives.

---

* AUTHOR'S "CHAUVINIST NOTE": At first, this may sound like a contradiction. Now I'm suggesting changing the site to suit the plans. But the end result will alter the beauty and ecology of the site less than most of the excuses for architecture which we see disturbing the landscape. Besides, none of the rules in this book should be taken too rigidly. I find it easy to bend the rules a little for earth-sheltered houses and beautiful girls!

Figure 54. *Earthwood, an earth-sheltered cordwood masonry structure.*

Living Systems

Planning rule number four: Consider *all* the living systems at the design stage, not just *shelter*. While we're going to all this trouble to put a roof over our heads, why not address the next most expensive "necessaries" of life, *food* and *fuel*? And what about *recreation*? And *home industry*? We, of course, are talking about the concept of whole living systems. Jaki and I are still working toward this concept, and paid particular attention to it in the design of our new house, Earthwood. I will use this design to illustrate a few of the "whole systems" considerations. Figure 54 is a perspective view of the design, Figure 55 the floor plan, Figure 56 the elevational plan.

ENERGY

Our new home, Earthwood, attends to low energy consumption through its design, orientation and various energy systems. The shape is cylindrical, the advantages of which have already been discussed. The fact that it is a two-storey house, however, also contributes to the ease of heating. If 2,000 square feet were spread over one floor, it would require some mechanical heating system to deliver the heat to all parts of the house. In Earthwood, no part of the house will be more than 16 feet from the heat source, a masonry stove 5 feet in diameter, discussed below. If rooms do not need to be kept to 70°F.—workrooms and bedrooms, for example, which we prefer at about 63°—the internal wall construction and the opening and closing of internal doors regulate the temperature in these areas. We do this now at Log End Cave.

The external walls are 40 percent earth-sheltered, virtually all of this occurring in the northern "hemisphere." Although the structure is earth-roofed, it is not a true underground house, but earth berming is almost as energy-efficient as a fully recessed building.[30]

The structural materials themselves contribute to a steady internal temperature. The walls, 16 inches thick, are of cordwood masonry. (Below grade, the cordwood is protected by an externally applied plaster layer and W. R. Grace and Co.'s Bituthene® waterproofing membrane.) The mortared portion of the wall consists of an inner and an outer 5-inch mortar joint, with six inches of fiberglass or lime-treated sawdust as insulation in the cavity.

These massive walls have a high insulative value or rate of about R-16, but of equal importance is their tremendous capacity to store heat for use when it is needed. Each cubic foot of cordwood masonry weighs about 30 pounds, so the walls of Earthwood weigh something like 70,000 pounds, or 35 tons. This is a tremendous mass for heat storage, a kind of internal thermal flywheel.

In addition, the centrally located masonry stove weighs 25 tons, and the 4-inch concrete floor (insulated with an inch of Dow Styrofoam® beneath) also weighs about 25 tons. (The earth roof weighs another 80 tons, but this is not very useful for winter heat storage because of its location above the Styrofoam® insulation. It is very useful, however, in keeping the house cool in summer.)

AUTHOR'S NOTE AS OF JULY, 1981:
During construction at Earthwood, we have uncovered a problem which can occur if too-dry wood is employed in cordwood masonry, especially hardwood. Too-dry wood can absorb moisture from the mortar, rain and/or high humidity and actually swell, causing stress cracking in the mortar joint. This is particularly serious on a curved wall, as the wall will tilt outwards, losing its plumb. We have not encountered the problem where we are using very dry cedar log-ends, which have enough internal air space to absorb the swelling wood fibers. As we do not have enough cedar to build the entire house, we are switching to surface-bonded blocks below grade, retaining the cedar log-ends above grade.

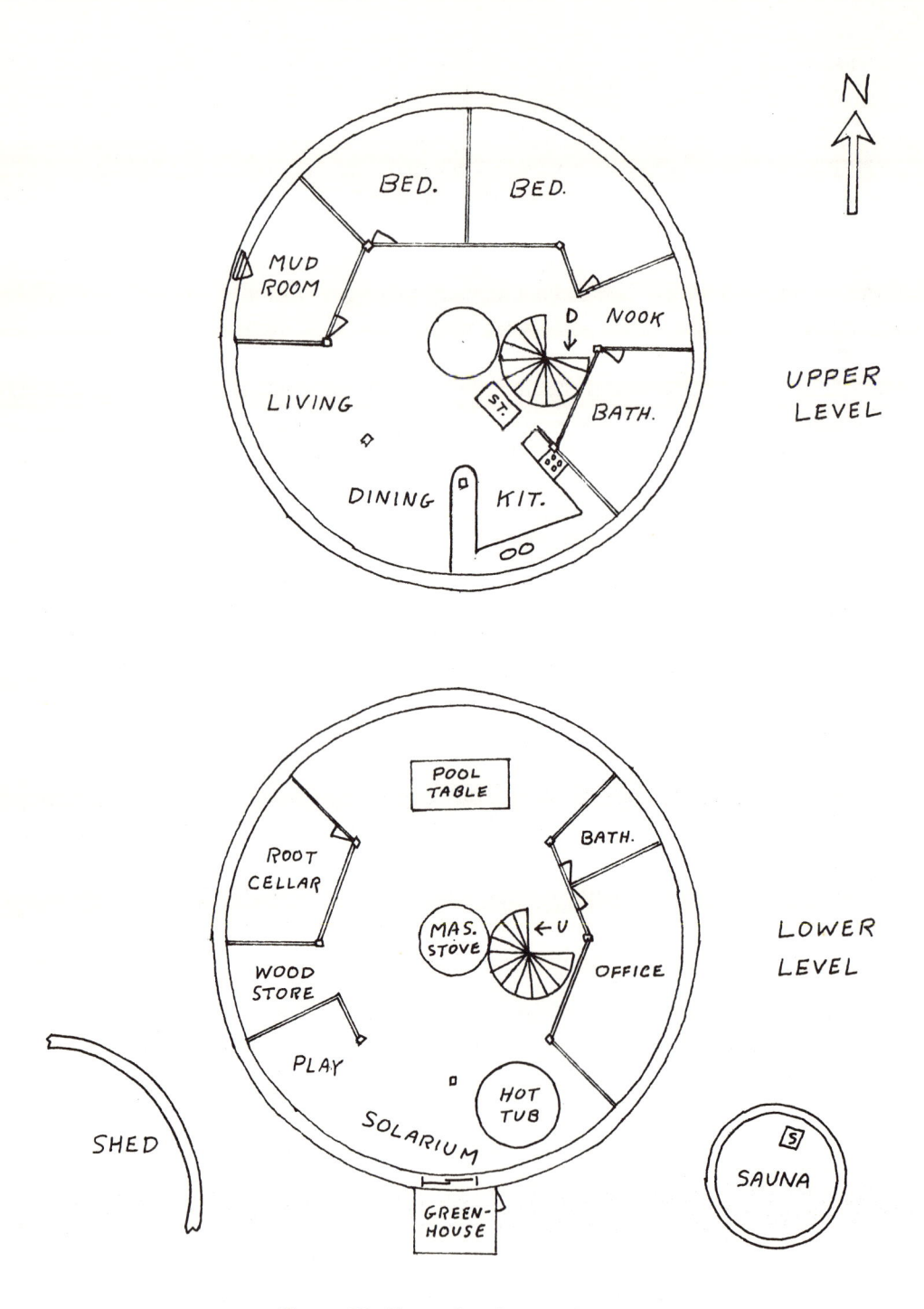

Figure 55. Floor plan for *Earthwood*.

How does thermal mass work? Consider the log-end walls of the Earthwood design and the following table:[31]

Material	Density lbs./cu. ft.	Specific Heat*	Storage Capacity**
Water	62	1.00	62.5
Iron	490	0.12	59
Concrete	140	0.23	32
Stone	170	0.21	36
Adobe	100	0.22	22
Sheetrock	50	0.27	13
Wood	30–40	0.30	9–12

* Btu/lb./°F.
** Btu/cu. ft./°F.

NOTE: Btu = heat required to raise one pound of water one degree Fahrenheit.

To give an idea of the comparative value of the British thermal unit (Btu), a 3-kilowatt electric heater produces just over 10,000 Btu's per hour. A 20-pound load of air-dried wood will produce about 140,000 Btu's,[32] but a percentage of this—50 percent in good stoves, 90 percent in the average fireplace—will be lost up the chimney.

It would require approximately 21,000 Btu's of heat to increase the cordwood wall temperature 1°F. (70,000 pounds × .30 specific heat × 1°F. = 21,000 Btu's.) Similarly, 21,000 Btu's would have to be lost to drop the wall temperature by one degree. If infiltration is kept reasonably under control, the internal air will seek the temperature level of the walls, floor and ceiling. Insulation placed properly to the *outside* of the thermal mass directs the stored heat back into the house, instead of to the outside.* Bringing down the wall temperature from 70°F. to 60°F. would require that 210,000 Btu's of heat be given up. The stone temperature in the masonry stove, once it had dropped to 70°F., would have to give off 105,000 Btu's to drop 10 degrees further, as would the 4-inch concrete floor. A hot tub, 6 feet in diameter and 3 feet deep, will store another 53,000 Btu's within this temperature range.

---

* AUTHOR' NOTE: This works very well on that portion of the cordwood wall below grade, where dense hardwood is used and the externally-applied Styrofoam® slows the heat transfer to the earth. Above grade, we will use white cedar, which has a much higher insulative value, but less mass. Because wood has both insulative and heat storage characteristics, it is difficult to arrive at even theoretical heat retention figures for above-grade cordwood walls. While the insulative value has been demonstrated to be about R1 per inch of wall thickness, it is also true that stored heat in the log-ends themselves will seek some level between the inside and outside temperatures. The inner mortar joint, however, is a very useful thermal mass in winter, just as the outer mortar joint has a moderating effect on summer heat.

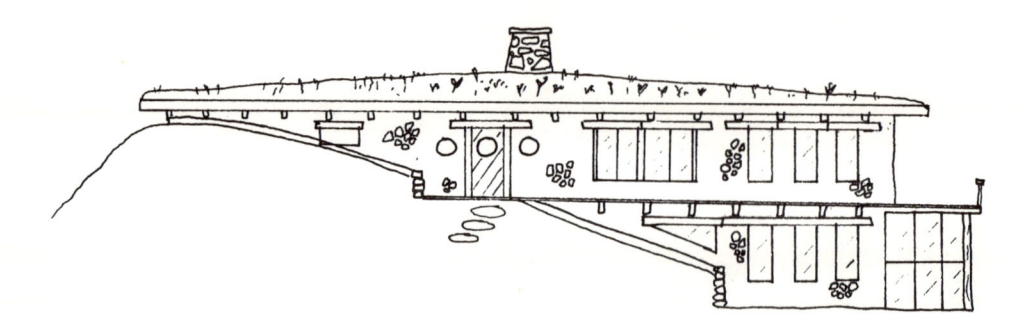

Figure 56. West elevation of *Earthwood*.

Internal walls, the ceiling and furnishings all help, too, but to a lesser degree, perhaps 50,000 Btu's to drop 10°F.

All told, 523,000 Btu's would have to be given off by the thermal mass for the house temperature to drop 10°. The time it will take for this to occur will depend on the insulation, infiltration, outside temperature and other factors, but the point is that Earthwood is designed with several times the thermal mass of an ordinary stick-framed house, a great advantage in moderating rapid temperature change.

THE MASONRY STOVE

Eastern Europeans say that Americans don't know how to burn wood, and I'm inclined to agree with them. Most Americans know that fireplaces are hopelessly inefficient, and will actually empty the heat out of a house in the wee hours when the fire goes out, thanks to the draft established in the chimney flue. (This loss can be greatly reduced by the installation of airtight fireplace doors.)

Modern wood stoves are a tremendous improvement, of course, but there are still problems. In an effort to keep the fire alive overnight, the airtight dampers are closed and combustion slows down greatly. Unfortunately, the firebox temperature drops below that required for complete combustion, causing unburned volatile gases to start their way up the chimney where—even more unfortunately—they condense in the form of creosote. Dirty and dangerous. Enter the masonry stove, also known as a "Russian fireplace" (Figure 57.)

It works like this. The stove is fired up once a day—the right time will be determined by experimentation—and an armload of wood is allowed to burn extra hot; that is, with plenty of air coming through the damper control. Perhaps a second charge of wood is loaded after an hour or two, perhaps not —this depends on the time of year. While the wood is being burned at high temperature, all the volatiles are combusted and there is no creosote buildup.

When all the yellow flames are gone, and only blue or nearly-clear flames are coming off the hot coals, the blast gate is closed and the fire is allowed to

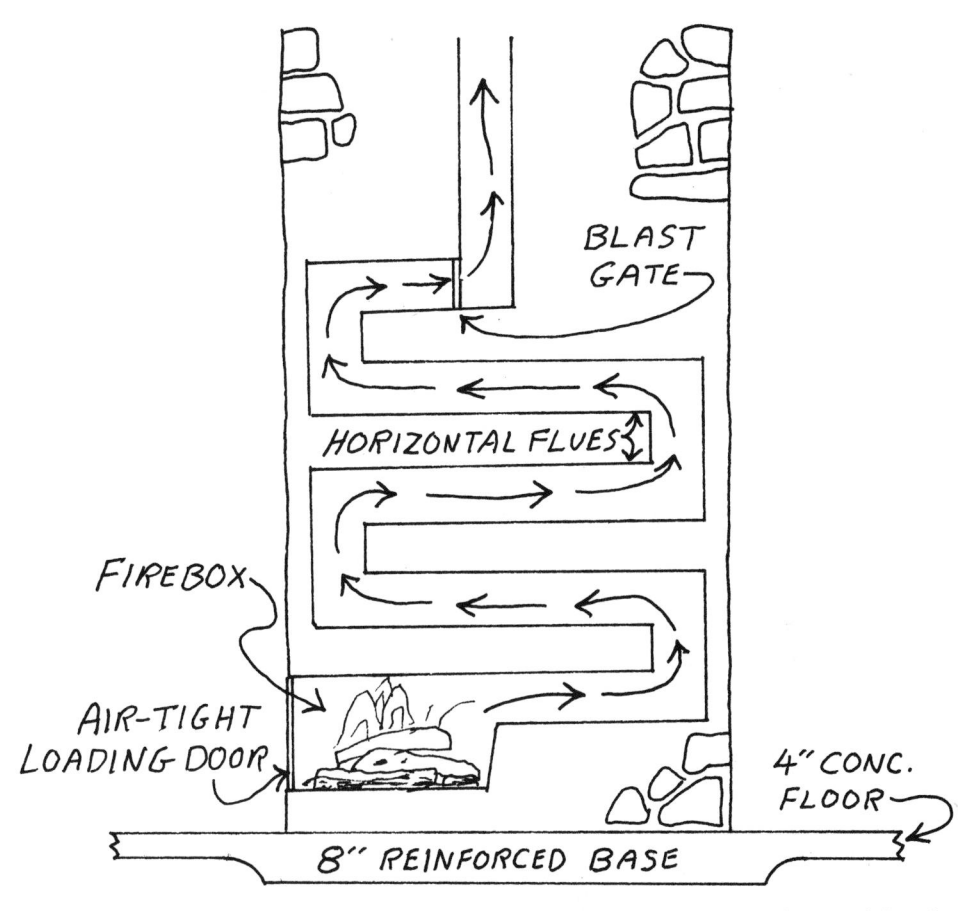

Figure 57. The heat normally lost out of a straight chimney is absorbed by the mass in a masonry stove, due to the increased surface area of the flue liner. When only hot coals are left, the blast gate is closed.

transfer its heat to the stone mass in its own good time. The extremely long flues, criss-crossing throughout the masonry, provide lots of surface area for the transfer of the exhaust heat to the thermal mass. If properly designed and built, the exhaust into the atmosphere is cooled enough so that you can hold your hand comfortably over the chimney top—*and* virtually no pollutants are put into the atmosphere.

The tremendous mass of the stove stores the heat, giving it off slowly to the house interior. Twenty-four hours later, the mass is recharged with a new fire. With masonry stoves, less fuel is used, the fire burns cleaner (little ash is left) and the temperature is steady. The only drawback is that more kindling wood must be stocked. However, this is offset by the advantage of having to fire the stove and clean the firebox much less frequently.

The hot tub referred to will be heated by a pipe passing through the heart of the masonry stove, with propane gas and—possibly—solar backup systems. A relaxing place to soak after a sauna or hard physical work or play, the hot

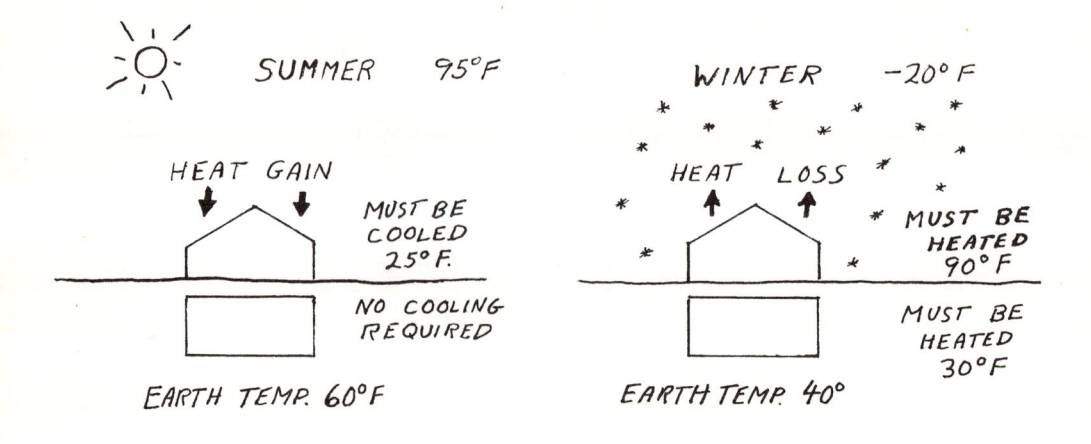

Figure 58. The heating and cooling advantages of a subterranean house.

tub also will serve as mass heat storage and a humidifier, which can be adjusted with a removable cover. (Cordwood walls promote a very dry internal atmosphere, especially in combination with wood heat, so the option of natural humidification may prove to be very valuable.)

EARTH SHELTERING

Look at Figure 58. An above-ground structure is situated in ambient (surrounding) air, which commonly reaches −20°F. in our part of the world. If 70°F. is required in the house, it is necessary to heat to 90° above the ambient temperature. In an earth-sheltered house, the ambient is the earth itself, which reaches a low of about 40° during a north country winter, so it is necessary to exceed the ambient temperature by only 30°. The left side of the diagram shows how a similar advantage is realized in summer cooling.

Log End Cave's walls are about 75 percent bermed. At Earthwood, this will drop to 40 percent, but virtually all of this will be on the north side. Improved insulation, the round two-storey design and the use of the masonry stove as the primary heat source will more than make up for the decreased use of earth-sheltering. Log End Cave, with 910 usable square feet of living space, requires 3 cords of wood per year for heating, about 300 square feet of heated area per cord. My estimate for Earthwood is that 4 cords of wood will be needed to heat the 2,000 usable square feet, or 500 square feet per cord.

WINDOW PLACEMENT

Windows should be *concentrated* on the *south* side of a house to promote winter heat gain and *minimized* on the *north* side to reduce heat loss. While Earthwood does not truly have "sides," it can be seen that the thermopane windows are concentrated in the southern hemisphere. A generous overhang protects the house from too much heat gain in summer, while permitting admittance of the low winter rays.

A solar greenhouse encloses the lower story entrance and will assist in heating on sunny winter days. Two triple-pane skylights admit additional light to the rear bedrooms and also to the stairwell area. Skylights are *not* energy-efficient, which is why I am now turning to triple-panes, but they are extremely light-efficient, admitting about five times the light of a comparably sized wall window. I consider a dark house to be a poor trade-off for, say, another quarter cord of firewood saved.

OTHER ENERGY

Upstairs "heat boosting" and cooking will be provided by an airtight oval cooking range. Electricity will be supplied by a Sencenbaugh 500-watt wind generator and deep-cycle battery system. This will be backed up by a yet undetermined amount of photovoltaic cells and a simple battery charger made from a lawn mower engine. For a discussion of consumption expectations with this sort of system, see *Wind* and *Solar Power* in Chapter 2.

Incidentally, electricity—even 12-volt electricity—should not be considered to be an absolute requirement in a home. Most of the people in our hill community—indeed, most of the people in the *world*—live quite happily without it. Bernard Rudofsky puts it rather succinctly:

> If we rate running hot water and electricity as *minimum* requirements for even the most humble home, Versailles and Windsor in their heyday would seem to have been just oversized hovels. Future generations may very well look back at our vaunted living standards as subhuman.[33]

FOOD

Food production and preservation should be a part of the construction design. At Earthwood, the greenhouse will be a part of the living unit, instead of a separate building, as at Log End. Ditto the root cellar. The garden will be sheltered from the northwest winds by the house and the garden shed. At Log End, the garden shed is a hundred feet from the house and the garden itself is another hundred feet beyond that. Close proximity to the house and the well mean easier watering of the raised-bed gardens.

We intend to use roof runoff to keep two large oak barrels filled with water. These will irrigate the garden beds via a watering system consisting of a plastic tube with capillary wicks extending from it to give a light but steady watering to the plants. This commercially available system—*Submatic* and *Tricklewick* are the brand names of two manufacturers—is powered by a gravity siphon. When the rain barrels are empty, they can be filled easily with a hand-pump situated directly above the well. (Even the hot tub can supply 600 gallons of lukewarm water to the garden when it is time to clean the tub. We won't be accused of wasting water!) All of this is possible by paying attention to the integration of systems at the time of designing the house.

Our root cellar will keep naturally cool, perfect for storing potatoes, onions, squash, apples and all canned goods. With its concrete block walls and concrete ceiling, this room would also serve as an effective fallout shelter, although I hope I never get to prove it.

RECREATION

Here's where I blow my "cover," showing that even the most practical designer will yield to temptation. Besides beer, *one* of my other vices—I'm not going to tell you *all* of them—is pocket billiards. This explains about 250 square feet of Earthwood's 2,000 square foot total. It also created an interesting structural problem. How do I eliminate one of the posts shown in Figure 50?

Remember the 140-pound-per-square-foot earth roof? A day's pencil pushing revealed that we would require a new 10-by-12 oak girder to span the 15-foot distance created by the elimination of one post. The darn thing will cost us $130 and weigh about 900 pounds, but I decided that this was preferable to a post rising up out of the center of the table. Giving in to my love of pocket billiards is costing us money in the form of a bigger house and an expensive beam, and trouble in the form of transporting and raising this monolith. No prize will be awarded to the reader who spots the most violations of low-cost strategies by this foolery. My excuse is that the house will be inexpensive enough to allow this calculated extravagance. (Materials cost estimate: about $8 per square foot, less than Log End Cave.)

Jaki is lukewarm about billiards, but is a terror at Ping-Pong, so we will be including a Ping-Pong table in the recreation room, too.

We've waited seven years for this bit of luxury, waited until we could afford it as well as reach a necessary level of building competence to go ahead. First-time owner-builders would be well-advised to complete the necessities first, adding the luxuries as they become affordable. And, hopefully, your recreational interests will require less than 250 square feet of living space.

HOME INDUSTRY

Ah, back to practicality! Back to work! Well, maybe not. Many people who build their own homes want to become free of the rat-race in other ways, too. Writers, artists, potters, private machinists, beekeepers—the list of home businesses is virtually open-ended.

Space for your home occupation should be planned very carefully. Some occupations can be conducted in a relatively small office within the main structure. Others may require an attached or unattached outbuilding. My 50-square-foot office at Log End Cave was adequate for a while, but now I find it woefully cramped. Earthwood will have an office of about 125 square feet, which should make it possible to carry on design work and writing without having to "clear the decks" each time I change from one to the other.

Home industry does not necessarily mean income-earning industry. Don't forget such homesteading activities as food drying, canning, beer making, sewing and workshop projects. Plan for them in the design, even if only in the form of a nonspecifically designated work area. Remember: work to *save* money, not to pay someone else to do what you can do yourself.

THE ECONOMY OF TWO STOREYS

As for Earthwood, I will not deny that the house is bigger than it absolutely needs to be. Readers of my book *Cordwood Masonry Houses* know that the Earthwood design is an evolution of the single-storey round house discussed in that book and illustrated in Figure 52 of this one. What finally persuaded me to go with the two-storey design instead, strange though it may seem, was economics.

I estimated that the original 1,000-square-foot plan, as outlined in the previous book, would cost about $12,000 for materials, or $12 per square foot. To add a second storey of the same size would only require about 5 extra cords of wood for the log-ends, a few extra posts, joists, 1,000 square feet of flooring, a stairway and some extra internal wall framing and covering, including a few more doors and windows. The foundation and earth roof structure, plus all the ancillary systems such as the septic tank, driveway access and energy systems would remain constant. I figured the cost of the second floor to be about $4,000, or just $4 per square foot. It seemed too cheap to turn down, and it did give a lot more scope for such things as a hot tub, a pool table and a reasonable office.

I mention all this not as an excuse for the size of the new house, but to illustrate that a two-storey house is more cost-effective per square foot than a single-storey house. Foundations and roofs are two of the costlier items in a structure, so let us minimize them in relation to the floor space. The sprawling "high school design" popular in the 1960's and 70's ought to be banned as outrageously expensive, both to build and to heat. Warm air rises. Let's use it twice before it leaves us.

Building Strategies

I've given a lot of space to planning strategies, because this is where the greatest savings can occur with the least trouble. Pencils and scrap paper are cheaper than rafters and sheathing. Without getting into specific styles of construction, there are certain general strategies which have proven themselves to be valuable to the owner-builder. Let's examine some of these.

LABOR BARTER

Two people can get more than twice the work done than can a single person. Except for individuals of rare tenacity and self-discipline, the solitary worker is easily discouraged, is prone to loneliness and laziness and has been known to throw his hands up in the air and walk off the site, never to return.

Two (or more) people keep each other going. This is not limited to husbands and wives. If two individuals or two families are building at the same

Figure 59. Log End Cave. This house was not built with expansion in mind.

time, as often happened in our hill community, it makes a lot of sense to trade workdays on each other's projects. More work will get done, and both parties will learn by the experience.

If you've asked someone over to help peel logs, for example, you'll be sure to get some logs peeled. The job isn't *that* enchanting that you can afford to waste an opportunity for some help. Also, certain jobs simply go faster with two or three helpers because of the mechanics involved, like frame raising.

Sometimes materials can be bartered with a neighbor who's building; sometimes materials can be bartered for labor. These deals are particularly economical if no money changes hands. They're *always* more satisfying spiritually, and less likely to backfire into bad feeling.

Work Parties

Occasionally, it will be advantageous to throw a work party at certain stages of the construction, particularly when it's time for the floor and footing pours, the wall raisings and rafter and roof work. On the other hand, I have seen owner/builders arrange for several friends—sometimes *too* many—to come over to help, and end up playing hosts, serving up a couple of cases of beer while the crew stands around talking. This lack of organization is the

Figure 60. Log End Cave's work party. The crew takes a well-earned rest after installing the third 30-foot 10-by-10 beam at the site.

fault of the owner/builder. He or she must be sure that all the required materials are ready the day before (it's too late on the morning of work), that jobs are properly organized, that the workers will bring the right tools for the job and that there are no pesky little details that have to be attended to before the day's work can begin. You can get a week's worth of work done in a day with organization. People come expecting and *wanting* to work. If they don't have a job waiting to be done, sure, they'll start in on the beer. *Might as well make a party of it. I've blown this day coming over for nothing anyway.*

Don't let this happen. Plan ahead—and involve *everyone*.

Part-Time Building

Many owner/builders build in their "spare" time, maintaining a regular job simultaneously, which provides the needed financial support to the project. Most of the case histories that follow illustrate this strategy.

In the Brasacchio's case, Frank kept his full-time job, working on the house as a second job, while Elizabeth worked on the house almost full time. Sometimes, a couple will take three or four years to build their house in this way, patiently staying in their present mortgaged or rented residence until the new place is completely finished.

Full-Time Building

This strategy means giving up your job, if you've got one. It requires a substantial grubstake or a high degree of skill and courage, like Ron and Debbie Light, whose story follows in the next chapter. The point is that the house gets built fast. This usually happens, but not always! Some people find that after eight hours at a desk, they are capable of putting in eight hours more at the end of a hammer, whereas they certainly would not be capable of putting in 16 straight hours on the house.

In general, if you are not relocating, part-time building is the safer strategy. When you are making a major move—perhaps for a change of life-style or to take advantage of more favorable land values—the full-time building approach is probably more viable. Get the house up as quickly as possible, and the saving on shelter costs will make up for the temporary loss of income.

Renovation

I am not eager to recommend old-house renovation as an economic strategy today. Buz and Jean Fitts, living in northern Vermont, think they might be among the last people of the decade to succeed at renovation as an economic strategy for owning a home.

They bought their large 19th-century farmhouse on its four-acre lot for $4,500 in 1975. (See Figure 61.) They live in the smaller wing attached to the back of the main structure. The large front part of the house is still several months from completion (March, 1981). Buz and Jean took out a loan to finance the project. When the place is truly finished, its value will be quite high, but it will be a bigger house than the Fitts family of four will ever require, unless they decide to open an inn, one of several ideas they have entertained. Another is to sell the house, perhaps keeping some of the land, and use the money to build a small, energy-efficient home.

Buz would be the first to tell you that building something new is a lot easier, and probably cheaper, than renovating an old structure in need of a lot of work. I would be the second. If the house is in good condition, it is not likely to be inexpensive. Old houses have become popular with the rich, pushing up their real estate values. Eleven-room Vermont farmhouses just don't go for $4,500 anymore. The work is slow and painstaking, and there are always complications: missing floorboards that have to be matched, a major girder to replace, a window or door frame than wants relocating.

If your intention is to retain the original character of the house, the project will be particularly expensive and time-consuming, as the exterior siding and the trim work will have to be matched, perhaps requiring custom work.

Sometimes, a small cabin can be found which can be whipped into livable condition quite quickly and won't add much to the cost of a parcel of land. Maybe the place just needs a good cleaning and fumigation. A situation like this could be quite valuable, as long as you don't spend too much time or money on it. Use the cabin as a temporary shelter; later it might prove useful

Figure 61. Buz and Jean Fitts renovated this 11-room Vermont farmhouse over a 6-year period, living in the annex to the rear of the building during renovation.

as a barn, storage area or workshop. Someday, when you've got lots of time and you're looking for a hobby, you might want to give it a complete restoration.

Kit Houses

Another option is to build a house from a kit. The appeal of working with a kit is that it frequently seems to offer the promise of saving money. Looking at the price lists of kits can create unrealistic expectations. A price quote of $20,000 for a 2,000-square-foot barn works out to $10 per square foot, a cost that hasn't been in effect since the late 1950's. You know that if you can count on family, friends and neighbors, labor won't add that much. You also know that the materials not included from the kit can often be purchased on sale from a salvage yard. At that rate, it would seem that a finished house shouldn't cost more than $30,000 or about half of today's usual prices.

While such calculations are common, they are overly optimistic. You can save money with house kits, but not 50 percent. The cost of a kit generally represents only 25 percent to 50 percent of the cost of the finished house. You must estimate the final cost very carefully or you may find yourself in for an unpleasant surprise.

My gut reaction is that kit houses will cost about twice as much as the comparable home which is entirely owner-built. On the plus side, a great deal of labor and decision-making will be avoided.

Organization

A common reason for problems and even failure among owner/builders is a lack of understanding as to the order of events in house construction and the priorities that result from this order. Even when the plans are fairly complete and sensible, well adapted to the site and the materials, it is still all too common that no thought has been given to the implementation of the plans. The basic order for virtually any kind of house is:

1. Site preparation
2. Foundation
3. Framing (or masonry walls and roof framing)
4. Roof
5. Plumbing and electric
6. Wall and floor finishing
7. Finish work

Every stage must be planned for in advance to assure that the necessary materials, tools and/or labor are not lacking. A few specific examples follow.

HEAVY EQUIPMENT CONTRACTORS

These contractors are in great demand as soon as the weather breaks in the spring. A logjam of house planners and contractors wait all winter to break loose. The whole project is stalled until the site work is done, so arrange for heavy equipment contracting as early as possible, preferably weeks ahead.

THE POUR

Prepare for your slab or footings pour a full 24 hours ahead of time! This may seem unduly cautious. It isn't. When you think you're ready, you'll still work a few hours checking over underfloor plumbing, form leveling, gathering the right tools, assembling the crew together, and so on. If you have a free hour to relax before the truck arrives, you'll have a more pleasant pour. Enjoy it. The next few hours are going to be the busiest time in the entire project!

MATERIALS

Lay by as many of the materials as you can prior to the building season. Spring also is the busiest season for sawmills, and you will have to build with green wood if you delay. Even a month's air drying of wood is a lot better than none at all. Do not predicate the whole project on a source of materials which is not absolutely reliable. Murphy's Law says you will be disappointed. I can personally vouch for this one.

PLUMBING AND ELECTRIC

If you are going to do your own plumbing and electric work—and you can and should if you are building the rest of the place—one of the best strategies is to find an experienced person to get you started, even if you have to hire him for a day to show you how to make the connections. Have a simple, easy-to-read plan and a couple of reference books handy.

Bathroom and kitchen should be adjacent for minimal plumbing. Avoid dimmers, three-way switches and the like; these complicate electrical work and add greatly to its cost.

Auctions and demolition sites are good places to look for recycled light fixtures, sinks, toilets and bathtubs. This can save you a small fortune. Remember that faucets are expensive and that an old sink with hardware is worth a lot more than one without fixtures. It is very difficult and expensive to find the right hardware for old sinks, toilets, and tubs.

WINTER

Get the place closed in before winter (if you live in the North). This is the only absolutely imperative deadline you have to meet. Winter will begin in your area a full two weeks before the earliest date you think possible. Of course, you lucky Southerners can ignore this.

FINISHING

This takes infinitely longer than you ever thought possible. Therefore, design for simple finishing. If you don't like sheetrock taping and painting, avoid sheetrock, or plan the joins to be covered with decorative wood stripping. I have no stomach for sheetrocking, wall-papering and window and door trimming, and will go out of my way at the design stage to eliminate these time-consuming features.

TIME ESTIMATION

It always takes longer to build a house than ever seemed possible. You observe a contractor putting up a house in five weeks and you say, "Well, then, surely I can build my house in five *months*." It would seem so. The professional cannot saw boards and hammer nails five times as fast as the amateur, but, because of his experience, he can think and make correct decisions five times as fast. In addition, he knows how to organize the job. Inaccurate time estimations often result in the owner/builder moving into the structure long before it is finished. *The Owner-Builder and the Code* makes the following well-founded observation:

> Besides cluttering the house with objects which must be moved or worked around, early occupation has the effect of altering the sequence of construction. This happens because there is an urgency to complete those aspects of the house which most affect the functions of daily living. The importance of such things as running water, counter space, lighting, privacy, and heating increases greatly when the house is occupied. An understanding of the principles of sequence is critical here, since owner-builders are suddenly faced with the prospect of installing several of the most complicated systems in the house.[34]

6

The People Who Did It!

The four parts of this chaper are *real* accounts of *real* people. Pay as much attention to the trials and tribulations as to the successes.

The first is the account of a single lady in her early 30s. Susan wrote her own "case history," just as it appears below.

SUSAN WARFORD

My homesteading plans evolved from a childhood fantasy. Whenever stresses built up, I dreamed of living in a house deep in a quiet woods. This house, of course, was comfortably equipped with necessities—dishwasher, clothes drier, modern tile bath, and so on. I never questioned whether I wanted central heat or hot running water. That was assumed. In fantasy nothing ever broke down, and I don't recall dreaming of utility bills, fuel shortages, mortgages or rising prices.

This vision was so rosy (or stresses so persistent) that I continued to dream of my retreat for years. I was 30 when it occurred to me, with jolting surprise, that I really could *live in a house in the woods if I wanted to. Ah, but how could I afford it? I couldn't buy a house, much less a forest. I had no savings and a below-average income.*

Luckily a friend introduced me to some homesteading literature. For her, it was amusing. For me, it was "salvation." Here was an enjoyable, ecologically sound, simple, practical way of building not only a house, but a way of living.

If I had the necessary funds, I might have rushed headlong into a homesteading venture. But, alas, I had no alternative but to wait while I accumulated cash. I began the exasperatingly slow process of paying off debts and charge accounts and saving, saving, saving.

But this waiting was a creative time, and ultimately I was grateful for the delay. I cranked up the old fantasy and began asking myself what homesteading activities I wanted to undertake. Hmmm—bees, goats, garden, orchard, wood heat, mushroom beds, a stone house.

Then I wondered how much land I'd need for these activities. Somewhere I had read that a cord of wood could be cut from an acre of mature forest each year without depleting it. And that a farmer used about 15 cords of wood to heat his house for the winter. Another source (my, but I was reading a lot) said that two goats could be pastured on an acre, and gardening books indicated that I could produce ample food on an acre. So it seemed that I'd need 15-20 acres, mostly wooded, to build my homestead. (I've since learned that the farmer must have been heating a huge drafty house. A small, well-insulated home can be kept warm on half the firewood he cut. So I needed much less land than I thought.)

I started a notebook on gardening and homesteading ideas, and in it I also made lists—of equipment, tools, contents of a first aid kit, books to read, clothing I'd need, goals, crops to grow, and so on.

I began gathering hand tools, kerosene lamps, ropes, buckets, mosquito netting, water cans, work gloves, compass, tarps, first aid supplies, and my closets bulged. I also gathered experience. I experimented with raising earthworms in my apartment, making yogurt, baking bread and candlemaking. I took a class in gardening and got hands-on practice by spending a week on a friend's homestead in Maryland and helping another friend prune grapes.

Whenever I traveled I looked at land. I bought newspapers for their farm and land ads, and I sometimes stopped to talk to real estate people. I visited several parcels of land in Ohio, where I lived. Eventually it became clear that I could afford $250/acre for land. It also became clear that there wasn't a lot of land available at that price, but there was some, mostly marginal. I looked at land well ahead of the time I was able to purchase, and this proved wise (but frustrating) because when a good buy appeared, I had an accurate idea of its value and could act quickly to get it.

At one point I wrote to a commune and looked into a land trust. I also designed a temporary shelter and made yet another list—of building materials, this time. Above all, I dreamed, and this had practical aspects, too, since I learned from my pipedreams about the kind of life I wanted to lead and the kind of person I am.

Late in 1974 I learned of Rob and Jaki Roy's proposed land division, and I began corresponding with them. In April of 1975, just after two and a half feet of snow blanketed the area, I spent two weeks slogging through drifts to see what I could of the land, and meeting the Roys and the other prospective homesteaders. Those who were interested in buying shared the cost of a title search and hired a surveyor who helped us rough out the boundaries (not an official survey). As I recall, my share of this expense was $50 . . .

Another major advantage of the group purchase was the creation of a supportive community of neighbors. This meant a lot to me since I was a woman alone and had never even built a bird house.

Figure 62. Susan's capped basement. She's planning to build a house on top.

We each bought parcels individually, and mine was 17½ acres. I paid $180/acre, for a total of $3,150. The down payment was $700, the balance to be paid in 5 yearly installments of $490 plus interest at 6 percent on the unpaid balance. Associated legal fees were $125, and my first tax bill was $33.57. (I think of those as the good old days.)

I moved to the land in May 1976. In a few days I gathered the materials for a temporary shelter. My parents joined me, and while my mother tended camp and cookfire, my father and I began building a small A-frame. [Figure 15.]

The framing is light enough to cause most carpenters to gag, but the structure has stood firmly for 5 years and will go another 5, I believe. The floor platform rests on concrete blocks at the corners and a post in the center of each end. The blocks rest on the ground, or are settled slightly into it, but they are not mortared or below frost-level, and the platform is not tied to them. The posts are set about 2 feet into the ground. They support the ridgepole and also a girder under the floor joists. The joists and studs are all 2-by-4-by-12's on 24-inch centers.

The floor is springy to walk on, but firm enough to support furniture, firewood and me. The dimensions of the A-frame are 12 feet by 16 feet at the base (including a 2-foot porch), and about 10½ feet high. In the peak I put a sleeping loft. Windows and doors were donated by friends. For roofing, burlap strips were nailed over the cracks between plywood sheets and covered with asphalt. Rubber sheeting was nailed over the roof peak, and the whole thing was smeared with an asphalt-based aluminum coating. The cost of the A-frame was about $350, and it was not insulated.

Construction went quickly. My father was only with me a week, and in that time we finished the floor, slanting walls and roofing. We would have accomplished more, but the weather was rainy all week, causing us to lose one day entirely and parts of a few more days. So we covered the front of the A-frame with plastic and part of the back with aluminum sheets, moved my gear inside, and waved good-bye at the end of a week.

I later finished the end walls myself, framing the windows and doors clumsily, my neophyte status very evident. It took me a few weeks to finish, but that's primarily because I began diverting my attention to other projects.

One thing did encourage me to finish the A-frame, however. Bears. They began visiting me nightly, attracted by my food. I banged pots and pans and lit my kerosene lamps, but I soon learned that hollering did more good. Finally, I got the house closed in, and the bears stopped their nocturnal visits, and I began sleeping better.

The rest of the summer was dismal. I depleted my funds and energies with too many projects—a garden too big to keep up, a well, a root cellar and a woodpile, all at once. I looked for a job, and I witnessed the groaning demise of my car. After settling with the mechanic, I had $2 left. Panic. Depression.

I applied for welfare, but it was about a month before I received help, and by then I had finally found a job. Meanwhile, I wallowed in defeat, mocked by half-done projects. I was so overwhelmed by failures and limitations that I didn't even do those things which could have been accomplished without any money at all—weeding the garden and gathering firewood, for example.

When I began my job, I found that the well and root cellar could not be finished before onset of winter. A month later my car died again, and this time it couldn't be revived. I couldn't get a car loan, I couldn't get to work, my house wasn't insulated, and snow was already on the ground. I took a bus back to Ohio, telling myself it was only for the winter.

I mention this grim experience because I learned from it, and perhaps you might, too. First, I learned how hard it is to homestead alone. A second pair of hands helps a lot, of course, but even more important is the emotional support and encouragement (or the prodding) of a partner. I also learned to give attention to one project at a time, my hardest lesson. And I learned a few things about self-discipline which didn't come easy.

That winter I dusted off the fantasy, regained my enthusiasm, and began saving again (ho hum). The next summer I bought a used pick-up and went back to the land just long enough to finish the well. I did a lot of stone work on it and discovered that I did not want to build a stone house after all. It is too hard and time consuming for a woman alone—or at least this woman.

By fall of 1978 I moved back again, installing a small cast-iron stove, insulating the A-frame with 3½-inch fiberglass batts and covering the floor with ½-inch layer of newspaper, ½-inch layer of homosote sheets and cheap linoleum. Total winterization expense, including stove, was $200. I bought firewood and settled in.

By Christmas I had a job, and two feet of snow had fallen. Daily chores included bringing in firewood, emptying the chamber pot and hauling water. Groceries, water and laundry were pulled to the house on a spiffy orange sled, and I kept a path open by tromping to the road with my snowshoes. Despite living in a cramped, primitive little house, winter was fun. And so was rent-free living. Now I was saving faster and living on the land at the same time. Nice!

As winter progressed, however, one problem became acute. The tiny stove heated the tiny cabin well, but it only held a tiny bit of firewood in its tiny belly. After 2½ hours, the fire went out, and the A-frame cooled quickly. Night after night I returned to bone-chilling inside temperatures (22° F.). Perishable food perished.

Starting a fire wasn't easy either, since my firewood outside was fairly unprotected and didn't have a chance to dry out before being fed to the stove (due to little storage space inside). It took an hour or two before the place got warm, so I had to set my alarm clock every two hours to tend the stove so I could sleep at all.

Not surprisingly, I began planning a bigger house and a bigger stove. I drew plans for a house which could be built in stages. The first stage would be a basement, which I could live in. This phase would be the most laborious and most expensive of the units, but I thought I could do it that year.

To finance the basement I took an extra job, typing four hours a day. Fine. Now I was working 12 hours a day and building a basement in my spare time.

In June 1979, I cleared the house site. Yep, it's in the woods. The land sloped to the south, allowing large windows on that side. In July a carpenter friend, Gerald Seguin, built the footings and slab, with a labor force of assorted neighbors, friends and me. In August I laid up blocks with friends but without mortar, and in September another friend and I surface-bonded the walls. (Rob describes this technique in detail in his book Underground Houses: How to Build a Low-Cost Home, *Sterling*) I found it an excellent method for amateur builders, since it is easily correctable and does not demand special skills.)

In October my parents visited, and I slyly put my father to work again. Gerald Seguin helped with advice and skill as we reinforced the walls and put the roof on. A friend made and installed the windows, and another made a fine airtight stove. One of the joys of my basement is the evidence (everywhere) of the contributions of my friends.

I kept careful records of expenses for the basement until I got to the final stretch of closing it in. Then things moved too quickly and I didn't keep records at all for the last month. However, I could tell by the hole in my bank account that I had spent a total of $5,500 on the basement. ($2,000 was borrowed and $3,500 was saved while I lived in the A-frame.) This included *supplies, occasional equipment rental, some labor, hiring a backhoe for the basement and a bulldozer for the driveway, some minimal plumbing (drains), but no septic tank. (Phase No. 2 consists of septic system, well and a real bathroom. Won't that be civilized?)*

Living in the basement has been a mixed experience. I had serious problems with condensation and humidity at first because it wasn't insulated. Moist air condensed on the cold ceiling and it rained inside. Finally I spread hay on the roof, and that insulated it well. I was comfortable from then on. Because of the large span of windows on the south side, the living space is light. This is particularly true in winter, of course, when there are no leaves on the trees. I love having plenty of space for storage, furniture, firewood or just swinging my arms.

When I think that I was working 12 hours a day and using all of my free time to build, it doesn't sound like fun. It sounds like something nobody could make me do. However, I thrived on the project. It was self-created, self-directed, and I delighted in seeing my house in the woods take shape at last.

PAT DUNIHO'S STUMP HOUSE (Figure 17)

After some years in Texas, graphic artist Pat Duniho returned in 1977 to the North Country of his youth, with a view to building a house in the woods. It had to be cheap, as he had no money and no job. He was the kind of fellow that bank managers have trouble looking in the eye, so he didn't bother going into any banks for help.

A great believer in the power of positive thinking, Pat soon landed a job with a local print shop and accepted brother Terry's invitation to stay in a spare room of his home. It was through Terry that Jaki and I first met Pat. Terry knew that we had a piece of land we wanted to sell and that Pat was looking for just such a piece. We soon struck a bargain, and became friends in the process, comparing notes on our common interests in land, building and self-reliance.

Pat is appalled by this "throwaway" society, and felt that the right shelter strategy should make use of waste materials. At work, he obtained a pile of aluminum printing plates which were going to be thrown out. Having seen the roof of our shed, he knew that the plates would be valuable when it came time to roof over his own house. He continued to keep his eyes and ears open throughout the building process and learned that it also helped to keep one's mouth open, too, at least part of the time, letting friends and acquaintances know of particular needs. Finding usable salvage is not luck, though it may appear to be so. Good scroungers work hard at their trade. Anyway, Pat obtained various articles and materials throughout the winter of 1978-79, some for free, some almost for free.

Pat lost his job during the spring of 1978 and, as the Roy economic situation was little better than Pat's at the time, Pat and I undertook to build a sundeck for the Terry Duniho family. One day, during a lunch break, we were discussing various structural possibilities for Pat's house. Being involved with a deck at the time, our thoughts turned to a house built on an oversize platform, part of which would remain as a pleasant deck in the woods. One of Pat's acquisitions during the past winter had been an old leaning silo of 2-by-6 tongue-and-groove planking, for which he'd paid $100. The silo yielded 2,000 board feet of good decking material.

We'd dug several holes in Terry's lawn for the pressure-treated timbers that were to hold up his deck and Pat expressed his reluctance to repeat the job through the roots and stones at his woodlot.

"Well, then," I said, "why not use tree stumps for posts? You can use the rest of the tree for a post and beam framework." I made the suggestion thinking that Pat was really after a low-cost temporary shelter to spend a few years in while accumulating the materials for a larger house. We decided that if the stumps were barked and treated, they'd last at least five years. Pat later learned that our idea was in no way new, that many Adirondack cabins had been built on stump foundations and that they would easily last 20 years or more.

Pat spent several hours looking over his land for the right clump of trees, and, when he'd found it—a tightly knit stand of 6-inch-to-8-inch-diameter balsam firs—plotted their locations on a chart. Armed with this information, he worked out the best rafter system and site orientation on paper. The house shape which evolved from this plan is trapezoidal, and the east-facing outdoor deck which remained is triangular. (Figure 63 is one of Pat's early pencil sketches of his house.)

Pat was beginning to take in a little money as a freelance commercial artist, and, with enough for gas and oil for his chainsaw, he began his "Stump House." (See Figures 64-67). His schedule allowed him to work three or four days each week on the foundation, cutting trees, barking the logs, barking and treating the stumps, fitting the first logs to the leveled stumps as floor joists, decking the whole thing with the silo boards.

During construction of the deck, Pat took advantage of hiring a backhoe which was already on the hill doing work for a neighbor. For $15, Pat's hole and ditches for his simple septic system were in place. By the time the foundation, deck and septic system were complete, Pat had invested $135 in the project.

Using straight balsam firs growing rather too close to the house, Pat cut the required posts and beams for his main frame. Neighbors helped raise the 14-foot tall south wall, and it was temporarily braced. Pat was able to finish the rest of the heavy framing himself with the help of a pulley.

The joists for the second storey were to be leftover 2-by-6's from the silo, but Pat lacked the roof rafters and let his need be known to anyone who might possibly have a useful suggestion. Before long, an acquaintance said that he had a large pile of wood in his backyard which he wanted removed. The pile was the remains of an old deck and roof which had been left to rot, although most of the material was still usable, including the rafters of the old roof structure.

Again, by word of mouth, Pat learned of a source of roof decking at trifling cost. A local moving company sometimes discarded old shipping crates, and Pat was able to pick up four very large ones for $60 the lot, containing a total

Figure 63. A preliminary drawing of the Stump House, by Pat Duniho.

of 24 sheets of 4 feet by 7 feet 6 inches by ¾-inch plywood and 58 2-by-4's. Dismantling the crates yielded a bucket of good nails.

With the welcomed help of another brother, Danny, the roof—consisting of two layers of plywood separated by a 1-inch dead air space—was finished in two days. Pat and Danny applied two layers of aluminum-faced builder's foil between the sheets of plywood, according to a plan found in Rex Roberts's *Your Engineered House* (1964, M. Evans and Co.). Pat believes that the foil reflects 80 percent of the radiant heat of the woodstove back into the room.

Pat used the same system of insulating the walls, which are made of 1,000 board feet of new 1-inch rough-cut lumber, costing $150. (Note: This material cost about $275 per 1,000 board feet in March of 1981.)

The windows, all strategically placed on the south side of the house, were donated by another friend. They originated in an old horse barn and had most of the multi-paned glass panels either broken or missing. Pat reconditioned the 10 windows in about 40 hours, at a material cost of $35.

Four months from the day he cut the first tree, the basic shell of the Stump House was up at a total cost of $500. It must be remembered that Pat worked

Figure 64. Barked, creosote-treated balsam fir stumps support the girder and floor joist system, made from the top logs.

mostly alone in the woods and only part-time. Also, he is, by his own admission, "a lazy person."

Two of Pat's acquisitions of the previous winter were old woodstoves, an air-leaking Franklin and a combination gas-wood cook stove. The combined cost was $90. With a can of furnace cement and some aluminum foil, he transformed the Franklin into an efficient wood-burner. The cement sealed the leaks at the joints, but the drafty front doors were the real problem. Pat plastered a generous amount of furnace cement around the door frame and, while still wet, covered it with aluminum foil. He then built a fire and closed the doors. Since cast iron expands when heated, he wanted the cement to harden while the stove was hot, to achieve the best fit. Pat now has an airtight Franklin—seemingly a contradiction in terms—that will hold a fire for 16 hours. As long as a small amount of radiant heat is being produced by the stove, the house is warm—despite the severe heat loss at night through his windows. Pat attributes the comfort to the radiant heat reflection off the builder's foil.

Besides foil and air spaces as insulation, Pat borrowed two other ideas from Rex Roberts. One is his front door, of which he says: "The door, like everything else in the house, is unconventional. I made it several inches larger than the door opening, and hung it on the outside of the house. Not only is this easier, but it provides a better seal than conventional doors. Wood expands and contracts with the seasons, so doors rarely fit properly in the winter, causing heat loss. My door fits like a freezer door. Expansion and contraction does not affect the air-tight seal."

Figure 65. The Stump House deck, as seen from below.

Figure 66. The deck is made from recycled silo planking. The clump of three white birch trees accents the integration of the Stump House in the woods. The white pipe leads to the septic tank.

The other Roberts idea which Pat employed was his ventilation system. Pat believes that windows are poorly placed for ventilation purposes, and so his windows are permanently fixed, again cutting way down on heat loss by infiltration. The ventilation system, involving high vents on the warm (south) side of the house, and low vents on the cool (north) side of the house, makes its own breeze, even on a still summer day. (For details, see *Your Engineered House*.)

When plumbing was finished (Pat uses a rainwater collection system with three oil drums as storage), counters and shelves installed, interior partitions and stairway built, the final cost of the Stump House was $1,000, or about $2 per square foot. The temporary shelter strategy kind of fell by the wayside as the house grew, and now Pat talks of expanding the house to the east.

Pat's house was not easy to build. There isn't a right angle in the place. But the basic idea of using tree stumps as a foundation has proven itself to be very successful. A large deck could be built in the woods by this plan, and a simple framed rectilinear structure could be built somewhere on the deck where it fits intelligently. After three years, the stumps show not the slightest sign of deterioration and the 20-year lifespan seems a reasonable expectation. If Pat lives 20 years in his house, his annual rent will average out to $50.

Figure 67. The post-and-beam frame supports the trapezoidal single-pitch roof.

Figure 68. The living area at the Stump House.

Figure 69. Pat's kitchen area.

FRANK AND ELIZABETH BRASACCHIO

Frank (29) and Elizabeth (33) Brasacchio are from New York City. During the early 1970s—the height of the "back to nature" movement which began in the late 1960s—they found their interests turning toward the country. They visited friends in rural areas and went hiking in the mountains whenever Frank could get away from his job as a cab driver. In 1972, they began to save their land grubstake, which culminated in the purchase of 36 acres near the Canadian border for $6,000 in 1974.

In early 1975, they moved to a small house near their land, where they exchanged caretaking duties for rent. They found that the location of their land—30 miles from the nearest population center—made employment opportunities slim indeed. There was the occasional substitute teaching job—Frank was a certified teacher—but there were no opportunities for regular employment in any of the rural schools. In desperation, Frank took a job as a janitor in Plattsburgh, 30 miles away. Traveling was conducive neither to family life nor to saving money, and, with the end of their caretaking arrangement in sight, the Brasacchios rented, in December, 1975, a farmhouse only 12 miles from Frank's job and 2½ miles from the land which Jaki and I had found two years earlier.

Unfortunately, within a few days of moving, Frank was laid off from work. Merry Christmas. Now, they were 18 miles from their land—with no job, a four-year-old daughter, a $100 a month rent, and only $44 a week unemployment to live on. (Food stamps stretched their buying power to about $75 per week.) Situation desperate? No.

By careful budgeting, the Brasacchios actually saved a little of their $44 each week! Still, this could not continue as a way of life, and they were becoming disenchanted with their land, which had only scrub growth for forest. Although they still craved the homesteading lifestyle, they considered giving up their dream more than once and they set September as the deadline, the point of return to city life or trying some other rural area.

Jaki and I had met Frank and Elizabeth during the previous summer, while we were building Log End Cottage. They were interested in our project, as it paralleled their own aspirations, although they had doubts about nonelectric living. They helped us with setting out floor joists and other jobs of that nature in exchange for learning these things. They had had no previous building experience and—as ours was limited to our little shed—it was pretty much of a "blind leading the blind" situation. Over the next year, we watched the Brasacchio's situation with interest.

In August of 1976, after seven long months of unemployment, several major events occurred in rapid-fire succession. First, Frank found a job as an apprentice optician in Plattsburgh. Next, Elizabeth gave birth to their second child, a boy. Then, they sold their property for about the same money they had paid for it, financing the mortgage themselves. Finally, overcoming their fear of nonelectric living, they bought 18 acres across the road from us on the hill. This piece of land was not taken in our original land division, and

Jaki and I had been carrying the payments on it. The new land had an excellent woodlot, a good meadow for gardening and fruit trees, and cost $1,000 less than their previous parcel, although it was only half as big. Frank and Elizabeth decided that the quality of land is more important than the quantity.

Selling the other land financed the down payment on the new parcel and got their house-building grubstake off to a good start. Subsequent land payments were made from Frank's salary, which, though nothing spectacular, seemed so compared with $44 a week. The couple rapidly saved money for the building of their house.

That fall, they tried their first solo building project, a small outhouse. It was not quite plumb, but plumb enough to function properly. This turned out to be an excellent strategy. Not only was some building experience gained, but, when they came to build their house the following spring, the hellish North Country black fly and mosquito seasons were far more bearable. They spent the winter of 1976-77 planning their house, talking with experienced builders and studying books on frame-house construction, especially *Your Engineered House* by Rex Roberts, and L.O. Anderson's *Wood-Frame House Construction* (Sterling Publishing Co., Inc., Two Park Avenue, New York, NY 10016).

On April 15, 1977, Frank and Elizabeth literally broke ground. Their house was to be built on 12 pressure-treated 6-by-6 timbers. The holes for these were dug by hand to frost level, a 6-inch layer of gravel was placed at the bottom of each hole, as well as a 16-inch-by-16-inch-by-8-inch concrete footing. The timbers were set, plumbed and jammed in place with stones. The holes were filled with sand for drainage, carefully tamped.

It is a common fear among inexperienced builders that wooden piers, even if pressure-treated, will not last long. Cornerstones Building School founder Charles Wing disagrees, saying, "A conservative figure for the life of an 8-inch diameter pressure-treated pole under the worst conditions is 75 years."[35] Also, many towns classify houses built on posts as "seasonal" and tax them very much less, even though the quality and energy-efficiency of the house is very high.

The main beams spanning from pier to pier were made of two 10-by-10's nailed together. From this point on, construction was standard timber framing, except that local rough-cut wood was used instead of finished lumber. Not only did this cut the cost of the house by about a third, but the aesthetics were greatly improved. (See 70.) The exposed 2-by-8 joists have a warmth and character sadly lacking in typically smooth pieces.

Frank was working a full 5-day week, starting at either 10:00 A.M. or 1:00 P.M., depending on his schedule. Because they lived only 2½ uncongested miles away, Frank and Elizabeth could zip up the hill at 7:00 A.M. and get a couple of hours of work in before Frank had to leave, or a full morning on the "late" days. On his "early" days, in June and July, Frank was even able to put in a couple of hours of building after work.

Elizabeth stayed at the land with the kids and worked on the house while Frank was away. I remember the morning that they started the sub-flooring.

Frank, building correctly "to book," began laying the one-by subflooring planks at a 45° angle to the floor joists. He'd completed about 10 square feet in one corner when it was time to dash to his job.

"There, Babe, got the idea?" he said to Elizabeth. "Yeah, don't worry," came the reply. When Frank returned from work that evening, the subfloor was just about finished.

Every job was tackled with pretty much the same strategy. Frank, who had done most of the research, would take responsibility for the technique to be employed and would figure out how to get the job underway. Upon return from work—behold!—the little elves had completed the job. What a system!

On Frank's days off, they would really fly, although they were always careful to take one day off from building each week, a wise plan. Too many days without a break and the work ceases to be fun. Enthusiasm wanes. The project drags on.

On August 24th, the family moved into their new home. They had spent $4,000 and had worked 19 weeks, Frank "part-time," although I am sure this amounted to 20 or 25 hours per week. By the time the house was completely finished, a few months later, the cost had run to $5,927.42. (They kept track.) This figure includes a well and code-approved septic system, and amounts to about $7.50 per square foot for their original 800-square-foot house.

In 1981, they completed a 400-square-foot addition, using the same construction techniques and time strategies. All told, they have invested about $10,000 in their house, $5,000 in their land and à few hundred dollars in three outbuildings. Except for a monthly car payment, they are free of debt. From the time that they made their decision to move to the country—starting with no savings—until they actually moved into their own home there was a span of about 5½ years. Now, 9 years from "Day One," their homestead grows by the season in its vibrant youth: the maples are tapped, the asparagus spears ring in the beginning of the growing season, the fruit trees and berries begin to yield their lifelong bounty. There is a quiet excitement going on.

What would they emphasize to others just starting out? They were in vociferous agreement on this: the great benefit derived when two people work on the construction together. Building is a psychological strain. Two people will get more than twice the work done as a single person, who is more prone to discouragement and depression. Any able-bodied person can learn simple construction techniques. Frank and Elizabeth started out equally inexperienced, so they learned together. If one partner starts with some building experience, he or she should encourage the other to pitch in wherever possible, measuring, marking, and so on. Soon each will be sawing, nailing and making perceptive suggestions. The house will be truly *theirs*.

Living without fuel and electric bills is every bit as pleasant as living without mortgage payments, although some people may have difficulty envisioning themselves without their clothes washer, freezer and 26-inch color television. Frank and Elizabeth have found acceptable alternatives to the conveniences and luxuries which depend on 110-volt a.c. power. For years they used gas

Figure 70. Frank and Elizabeth Brasacchio's framed house is built on pressure-treated posts. The board-and-batten siding is rough-cut one-by lumber that they purchased at a local sawmill.

and kerosene lights, but are now changing over gradually to the 12-volt power system described below. Their refrigerator runs on propane gas. Their clothes washer is a hand-powered model, made by Jamesway. They can and store their surplus garden produce. Water is hand-pumped into the house from their own well. They heat, of course, entirely with wood, although the house is orientated to take advantage of the solar gain in winter.

For years the Brasacchios have used a 12-inch black-and-white television powered off a car battery. Now, they have expanded their system with wiring for 12-volt power all over the house. Commuting the 12 miles to work each day enables Frank to charge a second deep cycle battery in addition to his regular car battery. Upon returning home, Frank plugs his house battery into a socket on a post connected to the house by an underground cable—sort of like tying one's horse to the hitching post. Now they have 12-volt lights and a cassette stereo in addition to the TV. It is surprising how much power is available from a heavy-duty battery. For example, if the battery has 120 amp-hours of storage capacity, it could run a single 25-watt bulb with its 2 amps per hour draw for almost 60 hours, or 6 hours a night for 10 nights. At this point, the battery would be discharged, but a battery made for deep cycling is easily recharged without damage.

Figure 71. The new living room addition to the Brasacchio residence.

If you've ever had your car battery killed in a few hours because headlights were left on, you may be skeptical about running a house on 12-volt power. The explanation is that (1) car headlights have a very high power consumption, several times that of a 12-inch 12-volt black-and-white TV, for example, and (2) ordinary car batteries have a relatively low storage capacity and cannot be deep-cycled without damage.

Perhaps you're not impressed with the prospect of running one or two 25-watt lights, but it should be reiterated that at 12 volts direct current, this is equivalent to a 50- or 60-watt bulb at 110 volts a.c. Twelve-volt lights are vastly superior to gas or kerosene alternatives, in convenience, safety, light quality and economy. A complete discussion of using a car's alternator to charge a second battery for home use is found in *How To Be Your Own Power Company* (formerly *The Wilderness Home Power System,* Jim Cullen, 1978, 1980, Van Nostrand Reinhold Co.), which the Brasacchios found to be very valuable in showing them how to install their own system.

The Brasacchio house would harmonize with other well-built framed houses in town. "Ordinary" electric wiring could have been installed quite easily, although this probably would have added $1,000 to the final cost.

Figure 72. Elizabeth's kitchen is in the original part of the house.

RON AND DEBBIE LIGHT

(House, Figure 73)

Ron and Debbie Light moved to the hill during the early summer of 1976, primarily to help brother John build a house. John, a single man at the time, finished the foundation and moved back to New Jersey, but Ron and Debbie stayed on, living in a tent for the summer with their five-year-old boy and baby girl. In July, their modest grubstake of $400 accidentally fell out of Debbie's handbag.

So they started their adventure with no job, no money and a tent for shelter. On the positive side, Debbie's sister and family lived in a log house on the adjoining property, brother John was happy to have Ron and Deb build on his land and Ron had a streak of optimism that kept a smile on his face through the toughest times.

John Light, the Dorresteyns (the Lights' in-laws) and Rich and Anne McIntosh had originally bought 70 acres together in 1975 and divided it three ways. The land was just over $100 an acre, cheap enough for good land, although there was poor access along an unimproved dirt road. Years later, the Light brothers finally got around to dividing John's land between them, and, in 1980, John returned to the hill with a wife and little girl and put a house on the 1976 foundation.

But back to late summer, 1976. Armed with a chainsaw, and spurred on by the hard reality of four people living in a tent, Ron began to build a house for his family out of poplar (quaking aspen) logs. Why poplar? Well, the other trees on the property were dense hardwoods of lower value as insulation and higher value as firewood or—as in the case of maple—as sugar bush. The logs were heavy, but so is Ron's back, and course by course, the 12-foot by 16-foot two-storey cabin began to rise from the woodscape, supported by a simple foundation of flat stones.

Odd jobs provided just enough money for gas and food stamps. In exchange for helping a friend dismantle a barn, Ron shared in some of the salvaged material, including roof planking boards filled with bent nails which he carefully pulled and straightened.

"Had to," says Ron. "We couldn't afford new ones."

The Lights' tin roof was recycled from another building in town which was undergoing extensive remodeling. By now, cold weather had set in and tent living was out of the question. The Lights moved in with the Dorresteyn family for the winter. The cabin was closed in against the elements and Ron worked away on the interior throughout the winter, warmed—or partially warmed—by the heat from a borrowed stove which ate wood voraciously. The cabin, still without insulation, was an effective atmospheric radiator.

In March 1977, the Lights moved in. The only expenses in the home turned out to be insulation, specialty nails and a few other incidentals. The original 400-square-foot building cost about $400, a dollar per square foot.

In April, Ron landed a job at the nearby Miner Center, an agricultural research institute connected with Cornell University. Ron's job mainly in-

Figure 73. Ron and Debbie Light's three-stage house in the woods was built for less than $2,500, but it has served them well.

Figure 74. The south wall of the Light house, showing the original Poplar Palace (on right) and the cedar addition.

volved the experimental dairy herd, but he would float to other parts of the institute as required, gaining a knowledge of what was going on in the forestry department, nursery, and so on. Even though the job paid little more than minimum wage, Ron found that the side benefits of working there were valuable to his homebuilding and farmsteading activities. For example, Ron carried home piles of manure from the dairy barn, giving their raised bed gardens an incredibly rich start in life. Extra "experimental" apple trees formed the nucleus of a future orchard.

Although glad to be in their own home and free of debt at last, the Lights found their "poplar palace" rather cramped for a growing family. After considering the pros and cons of adding on or building anew (using the palace as a temporary shelter), the add-on strategy won the day.

During the summer of 1977, Ron was able to purchase large cedar trees from the Miner Center at a dollar each. He took 50 trees. With Debbie hefting her fair share, the Lights hauled them to and from a local sawmill, where they were milled two sides for logs and rafters, with the "scrap" made into 1-inch boarding. The project even yielded a good supply of prime cedar fence posts. The bulk of the materials for the addition, then, cost $110 plus gas for the truck, $60 of the cash going to the sawyer. The labor cost, however, was quite high. It took Ron and Debbie most of the summer to get the wood to the site in its final form, but this was acceptable as they still had more time than money.

Working with less pressure and more experience, Ron built the second part of their home to a much higher standard. The logs are vastly superior in energy-efficiency and appearance to those used in the original cabin, and so

Figure 75. These stairs join the four different levels at the Light house.

is the foundation, which consists of seven concrete pillars. Below grade, the hand-dug holes served as the forming; above grade, Ron used cardboard cylinders. Mixing his own concrete by hand, the foundation cost about $25 and is as stout, frostproof and permanent as it needs to be. Foundation work was also completed that summer, in whatever spare time remained.

The autumn of 1977 found the log work in the addition beginning to take shape. Because the addition had superior footings to the original structure, Ron incorporated a system of joining the logs of one building to the other, so that the original poplar logs can rise and fall in relationship to the new cedar logs via a keyway system made of 2-by-4's and stuffed with fiberglass. "The two sections have independent suspension," says Ron. It works.

Because of priorities, lack of extra money and the fact that Ron was working virtually alone and in his spare time, the home was just nearing completion before Christmas a year later. Ron worked the night of December 23rd right through until dawn, cutting the opening through the common log wall with a chainsaw at 6:00 A.M. After connecting a new stove and eating a well-deserved breakfast, Ron was off to work. Christmas Day found the Lights enjoying a spanking new living room and a second bedroom. Their space had doubled. So had the quality of construction and the cost. The addition, of similar size to the poplar palace, cost a little more than twice as much to build. A $100 sliding-glass-door unit (used), upstairs framing, lots of insulation and other windows and incidentals brought the cost up, although it was still less than two months' mortgage payments by the average American first-time homeowner today.

Typical of the kind of cost-saving decisions which they made is a large 4-foot by 5-foot thermopane window which Ron bought for a dollar because the airtight seal was broken. Ron caulked the unit as best he could and placed it within the west wall. In the wintertime, the moisture between the ¼-inch-thick panes turns to beautiful Jack Frost designs. Ron reckons that the window's performance as a thermopane is virtually the same as if the air space was completely dry. At a dollar, the trade-off is greatly in the Lights' favor, no pun intended.

In 1980, Ron built yet another addition, a bathroom/laundry attached to the cedar module. Having been without a proper bathroom for several years, Ron and Debbie spared no expense. The fiberglass tub and shower enclosure alone cost as much as the poplar palace. Ron admits that he could have saved considerable time, money and space, had he not elected to build the bathroom *around* a sugar maple tree, standing happily between the bathroom and laundry sections.

"Sometimes I wonder about the wisdom of that," says Ron, "but you gotta do what you gotta do." The Light house, like Pat Duniho's, is full of such design features, which give these owner-built homes that special flavor not commonly found in contractor-built homes. Clear, cold spring water is delivered to the bathroom and kitchen by gravity through a long plastic pipe meandering through the woods. In winter, the water is allowed to flow constantly, so that the pipe does not freeze. The clean overflow is channeled to a nearby stream. In summer, the extra water irrigates the garden.

All told, the Lights' three-stage house cost about $2,400 for the 1,000 square feet of area. *And* two years of cramped living, scrimping, and hard work. Although there are things they wish they had done differently—"I designed the house after I built it," says Ron—they are very happy with their aesthetic, ecologically attuned and *mortgage-free* home.

7

Conclusions

The previous chapter shows that it is possible for single and married people alike to own their own homes. I hope you noticed that hard work—lots of it—and willingness to put up with a certain degree of hardship for a while were common to the four case histories presented. Small grubstakes and low incomes were also common to all. Someone earning $15,000 per year might have accomplished more in a shorter time—if he or she also had the tenacity displayed by the heroes of these studies.

Nowhere in this book do I mean to imply that there are any "free lunches" with regard to shelter cost. The people on the hill have not been unusually lucky; they have made their own luck and worked hard to make the best use of it. The Lights' loss of their entire grubstake was not lucky—it probably cost them the chance to move into the poplar palace that first winter—but strength of character and like-minded neighbors helped them through this adversity. Pat's windfall of 10 large windows may seem lucky, but most people would have rejected them as a pile of junk. They looked pretty bad when I saw them, and I wondered if they were worth the effort of restoration. Pat was unafraid of the task, although he calls himself lazy, and put many hours of hard work sanding and reglazing to restore the windows. Susan's failures, described with such feeling in her story, were—as *experience*—the raw materials of her later successes.

It is not likely that you will achieve all that you want from your shelter as fast as you would like to have it. Patience may be the single most important requirement to achieve economic freedom and true security, nct the 30-year "patience" imposed upon us by lending institutions to achieve a false security, but the simple, daily kind of patience coupled with a strong desire to achieve one's goals. Not all who attempt to win this freedom by building their own homes will succeed. I hope that this book has helped you to decide whether or not you are the kind of person who will start with the odds in your favor. If your heart's not really in it, the odds are overwhelmingly against you. Do not begin the project. I will not want to receive your correspondence. But if you feel that you can in no way afford to delegate the responsibility of your own shelter to others, then examine your assets, formulate a plan of strategies and stay with the plan, adjusting it as necessary with changing circumstances. The exact path is not so important, but know the general direction of your goal. If the train you are presently riding is going the wrong way, get off at the next station.

Since mortgage rates hit 13½ percent in November of 1980, lending institutions have been coming up with all sorts of schemes to encourage homebuyers to go into even greater debt. As I write now, in March 1981, the rate has hit 15 percent. Some of the "incentives" that I have heard of lately have been: (1) to give a lower rate if the buyer agrees to share in the capital gain when the property is eventually sold (the bank becomes a *permanent* partner in your home), (2) to restructure the formulae for determining borrowing qualifications (you can now go into even greater debt on the same salary); (3) floating interest rates (don't expect them to float down); and (4) making it easier to get a second mortgage much earlier in the game, possibly as soon as the house is bought (I have no intelligent comment on this.)

Do not be persuaded by these ruses. They are brought to you by the makers of the plastic credit card, the other great enslavement device of the money masters. Mortgages and credit cards are at least as serious a problem as drugs in this country, and probably more addicting. We *sign up* for our addiction.

No, I prefer Thoreau's commentary of 125 years ago. In the concluding chapter of *Walden,* he says:

> I have learned this, at least, by my experiment: that if one advances confidently in the direction of his dreams, and endeavors to live the life which he has imagined, he will meet with a success unexpected in common hours. . . . If you have built castles in the air, your work need not be lost; that is where they should be. Now put the foundations under them.[36]

Appendix One

Strategy Chart

HOW TO USE THIS CHART

Intersect your income with your total savings to determine the key number for the appropriate strategies. Example: If your income is $10,000 a year and you have already saved $4,000, your key number is 5. Read strategy list 5. Another example: You are unemployed, but have $2,000 in the bank. Read lists 1 and 4, the ones nearest to your situation. This chart is based on values in Upstate New York in 1981. Other parts of the country may be more or less expensive to live or build in.*

As time goes on and inflation takes its toll, salaries and savings of equivalent empiric value to the 1981 figures should be used. This chart and the accompanying strategies should be considered only a rough guide. Your final plan of action must be based on a realistic appraisal of your own abilities, self-confidence and personal circumstances.

INCOME → SAVINGS	NONE (OR VERY LOW)	$7000 – $20,000	OVER $20,000
$0	1	2	3
$3000 – $5000	4	5	6
OVER $10,000	7	8	9

*AUTHOR'S NOTE OF MARCH 19, 1981:

On the NBC Nightly News "Special Segment" last night, the topic was the high cost of housing. Despite my familiarity with the subject, I was shocked at the figures for California. The *average* house in the Golden State costs $112,000! A couple with a combined annual salary of $37,000 were living in a mobile home, saying they couldn't afford a house. They felt cheated that their American Dream had been shattered. I wished I had their address, so I could send them a copy of this book.

According to NBC News, *55 percent of take-home pay* in California goes toward shelter costs. I was flabbergasted. And my Californian readers are probably equally flabbergasted by the kinds of figures I use in this book. The figures may vary around the country, but the philosophy remains the same.

ONE

1. Carry on your merry way, as long as you do not expect others to support you. Only then are you truly free.
2. Join a religious commune or monastery.
3. Marry a rich spouse.
4. Get a job and move to List Two.
5. Find someone involved in building a house. Exchange labor for experience, and room and board. Keep eyes and ears open to opportunities.
6. Build an ultra-low-cost house on someone else's land (with permission). This is what Thoreau, and Ron and Debbie Light did, so you're in good company. Read Chapter 6 again.

COMMENT: Tomorrow is the first day of the rest of your life.

TWO

1. Begin to assemble your grubstake.
2. Look for land to get an idea of what you want and what price you can expect to pay for it.
3. When possible, put a down payment on land. Know where future payments are coming from.
4. Lack of savings might be due to undeveloped sense of economy. Read Chapter 3 again and the books on economy listed in the Bibliography.
5. Read, research, try to get some building experience.
6. Look at Susan's list of activities which can be done while saving the grubstake (Case History: Susan Warford).

THREE

1. Change economic philosophy. If you are unable to save money on your present income, you:
   A. Are paying too much for rent or mortgage. Change shelter situation.
   B. Have too many kids or one or two spoiled ones. I don't think I can help you.
   C. Have no sense of economy at all. Money runs through your fingers like dry sand. Your purchases depreciate rapidly in value, like snowmobiles, big cars and banana splits.
2. Read Chapter 3 again and the books on economy listed in the Bibliography.
3. Alternative: Make no change at all. Maybe you are already leading the life which suits you best.

FOUR

1. Buy some land, even if it must be "on time." (If buying land "on time," you'd better find a job to save for next year's payment.)
2. Make the move as is. Susan, Pat and the Lights started with less.
3. Use the temporary shelter strategy.
4. Grow food, cut wood for heat.
5. While living in the temporary shelter, gather the materials for the main house.
6. Try to find at least a part-time job.

FIVE

1. Find and buy land. If land is close to work, build a temporary shelter to accelerate the saving of the grubstake.
2. Read, research.
3. Keep your job. Build part-time. See Frank and Elizabeth's case history.
4. Use the add-on house strategy.

COMMENT: Giving up your job to build full-time is risky with this sort of grubstake. Living expenses can eat up the savings rapidly.

SIX

1. Buy land, or at least make a down payment.
2. If land is close to work, build a temporary shelter to accelerate saving of the grubstake.
3. If you've followed (1) and (2), you have good potential for laying up a substantial grubstake in short order, but convert dollar savings to building materials or tools whenever the right opportunities come along. Otherwise, inflation can be disastrous.

COMMENT: I suspect that many of you see yourselves in this list of strategies. The income is a very valuable asset and should not be given up lightly (or at all, if you're one of the minority who are truly happy at your work).

SEVEN

1. What are you waiting for? $10,000 is more than enough. With no job, you can relocate to where land is cheap and building inspectors are far and few between.
2. Buy land, build a temporary shelter, build the permanent house. If the land is bought "on time," know where future land payments are coming from.
3. Unless you are opting out altogether—I've no objections—have a care about where your future living is going to come from.

EIGHT

1. Find and buy land.
2. Read and research.
3. If unhappy with your job, leave it and build your house, starting with a temporary shelter to save on grubstake depletion. The temporary shelter can be added to, or a new house built on a pay-as-you-go basis.
4. If relocating, keep alert to employment opportunities right from the start. Ten grand can disappear faster than you think.

NINE

1. Buy your land.
2. Read and research.
3. Answer this question honestly: Are you happy at your work? If *yes,* stay with

it. Build in your spare time. The temporary shelter strategy is optional. Don't use it if you're comfortable where you are, your land is nearby, and you are still able to save a decent amount of money each year.

If *no,* make the move or find another job. You only get one life. You are in better economic shape than 90 percent of the owner/builders I've known. Your chances for success are very high, as you have already shown a sense of economy.

Appendix Two

Owner/Builder Schools

Owner/builder schools have been around for a few years now. Based on conversations with students in classes that I have conducted, as well as with students from other schools, it seems that the level of satisfaction with the courses is very high. Information about the schools and the courses offered can be obtained from the directors at the addresses below.

B. Allan Mackie School
 of Log Building
B. Allan Mackie, director
P.O. Box 1205
Prince George, BC
Canada V2L 4V3

Going Solar
John Stebbins, director
216 Canyon Acres Drive
Laguna Beach, CA 92651

Earthwood School of Cordwood
 Masonry and Earth-Sheltered
 Construction
Robert L. Roy, director
RR#1, Box 105
West Chazy, NY 12992

Northern Owner-Builder
Paul R. Hanke, director
Route 1
Plainfield, VT 05667

Owner-Builder School
Richard Owens, director
2000 S. 5th Street
Minneapolis, MN 55455

Yestermorrow
John Connell, director
Box 344
Warren, VT

Cornerstones, Wing School
 of Shelter Technology
Rick Karg, director
54 Cumberland St.
Brunswick, ME 04011

Heartwood Owner-Builder School
Elias Velonis, director
Johnson Road
Washington, MA 01235

Minnesota Trailbound Log
 Building Program
Ron Brodigan, director
3544½ Grand Avenue
Minneapolis, MN 55408

Owner-Builder Center
Blaire Abee, director
1824 Fourth Street
Berkeley, CA 94710

Shelter Institute
Pat Hennin, director
38 Center Street
Bath, ME 04530

Indigenous Materials Housing
 Institute
Jack Henstridge, director
Box H
Upper Gagetown, NB
Canada E0G 3E0

BIBLIOGRAPHY

HOW TO USE THIS BIBLIOGRAPHY

I have used most of the books in this Bibliography, and have found them to be valuable and inspirational learning aids. The uneconomic way to use this list is to run out and buy all the titles which appeal to you. The economic way is to start at the library, borrowing a few of the books which interest you each week or two. Remember that most libraries can obtain a book for you, even if it is not normally carried. Sometimes they will buy the book outright, sometimes they will borrow it from another library. After reading the books, buy only those you know you will be referring to again and again. If you find yourself spending much over $40 for building books, you are probably not making the best use of the library or your money.

THE CATEGORIES AND THE KEY

The books are arranged by category. Sometimes a book like Rex Roberts' *Your Engineered Home* would fit conveniently in several different categories; it is *philosophical, inspirational,* accents *timber framing* as a construction technique, but is probably best classified as *general,* giving you a good *overview* of building considerations. The numbers in parentheses at the end of the commentary—as: (1,2,7)—indicate other areas in which the book would be useful. I do not have personal experience with some of the works and have noted these with an asterisk (*). They are listed as a start-up source for the particular subject. An "N" in the key indicates that a review and access information appear in *The Next Whole Earth Catalog.* (See Category 3, Access.)

ANNOTATED BIBLIOGRAPHY, CATEGORIES

1. Economic philosophy
2. Inspiration
3. Access
4. Land
5. Building, overview
6. Timber framing
7. Earth sheltering
8. Logs
9. Masonry (stone, adobe, cordwood, surface bonding, masonry stoves)
10. Renovation
11. Kit houses
12. Food
13. Energy
14. Urban homesteading
15. Miscellaneous

1. ECONOMIC PHILOSOPHY

The first two books in this list, by the immortals Thoreau and Hesse, have done more to instill in me a sense of empiric economics than all the rest of my reading combined.

Walden, Henry David Thoreau, 1854. (New American Library, Inc., P.O. Box 120, 120 Woodbine, Bergenfield, NJ 07621) The first chapter, Economy, is all that needs to be read as far as low-cost living and building is concerned. It runs about 50 pages. Don't be surprised if you find yourself reading with equal pleasure the remaining 180-odd pages, mostly naturalistic observations, generously marinated in the unique Thoreauvian wit and philosophy. (N)

Siddhartha, Hermann Hesse, 1922. (Bantam Books, 333 Avenue of the Americas, New York, NY 10014) *Siddhartha,* a book of economics? Absurd. Yet it worked as such, and much more, too, during my grubstaking days in Scotland. Rereading it for my work on this book, after a lapse of several years, I find that it still weaves its magic. *Siddhartha* is a story of essential living, which is another way of saying empiric economics.

Possum Living, Dolly Freed, 1978. (Universe Books, 381 Park Avenue South, New York, NY 10016) The sub-title is almost a review: *How to Live Well Without a Job and with Almost No Money.* This book will help you save the grubstake. This excerpt tells you why you should:

> Owning your own home free and clear—that's the key to all the rest. Once you have your snug harbor, your safe base, all else comes easy. You can tell the rest of the world to go to hell if you want, once you own the roof over your head. I believe that some parents who are willing to scrimp and save to give their kid a college education would be doing the kid a better turn by giving him that money to buy a house instead. Once he realizes he doesn't have to worry about his future—once he has security and leisure to think about it, instead of having his future rammed down his throat—he'll make his own future. (*,N)

Ectopian Encyclopedia for the 80's, Ernest Callenbach, 1980. (And/Or Press, Box 2246, Berkeley, CA 94702) Stewart Brand, a genuine expert on this sort of thing, calls Callenbach's book, "The nuts and bolts of enlightened self-interest. The cheap transportation, housing, food, clothing, etc., which is better than most of the expensive stuff."[37] The book is based on Callenbach's previous title *Living Poor with Style,* which tells what it's all about. Another good grubstaking guide. (*,N)

2. INSPIRATION

The Prodigious Builders, Bernard Rudofsky, 1977. (Harcourt Brace Jovanovich, 757 Third Avenue, New York, NY 10017) Vernacular architecture, integrating site with structure, and making use of indigenous materials. Good recipe. (N)

The Rock is My Home, Werner Blaser, 1976. (Van Nostrand Reinhold, 7625 Empire Drive, Florence, KY 41042) Incredible use of the toughest of indigenous materials. If you are planning to build of rock, let this book charge up your imagination. (9)

Handmade Houses: A Guide to the Woodbutcher's Art, Art Boericke and Barry Shapiro, 1973. (A & W Publishers, 95 Madison Avenue, New York, NY 10016) (N)

The Craftsman Builder, Art Boericke and Barry Shapiro, 1977. (Simon and Schuster, Rockefeller Center, 1230 Avenue of the Americas, New York, NY 10020) These two books by Boericke and Shapiro have a common character and design. Each is composed of approximately 100 quality color plates showing details of houses moulded by "vernacular architects" (a fancy name for people who roll up their sleeves and build themselves a home, usually with recycled or indigenous materials and often with*out* plans any more complicated than what can be drawn on a hunk of birch bark.) Each plate illustrates that shelter can be as free—in every sense—as the builder wants it to be. These books unlocked my imagination, and they could do the same for you.

3. ACCESS

The Next Whole Earth Catalog: Access to Tools, edited by Stewart Brand, 1980. (POINT, P.O. Box 428, Sausalito, CA 94966) The "tools" listed are mostly books. A publisher cannot pay to have his books listed. There are no advertisements. The chief Soft Technology (includes Housing) reviewer is J. Baldwin, now a part of the New Alchemy Institute, well known for its work in the appropriate technologies field. Lloyd Kahn, of *Shelter* fame, and editor Stewart Brand also do a lot of reviews in the Soft Tech department. Their reviews are concise and to the point, backed with illustrations and excerpts from the books reviewed. I'm the kind of person who lets a man know when I disagree with him, but I have to say that—based on reviews of the books with which I have personal experience—I agree with these three gentlemen 99 percent of the time. Of all the books in this Bibliography, the *Catalog* is the one that I would recommend that you own, as it puts you in touch with sources of information that you are likely to need. And *they* didn't pay *me* to say *this,* either. Most of the books listed in the *Catalog* can be ordered either through the Whole Earth Household Store or from the publishers.
CoEvolution Quarterly, P.O. Box 428, Sausalito, CA 94966. Sort of an on-going *Whole Earth Catalog*—produced by the same people, too.
The Mother Earth News, P.O. Box 70, Hendersonville, NC 28791. This magazine is both a practical nuts-and-bolts type of homesteading journal and an access source for further information on alternatives subjects. "Mother's Bookshelf" contains over 700 titles, including many appearing in this Bibliography. Their catalog is free, same address.

4. LAND

Finding and Buying Your Place in the Country, Les Scher, 1974. (Macmillan Books, Front and Brown Streets, Riverside, NJ 08075) The only book you need on buying land. Covers everything: looking for land, evaluating the land, financing, legal, the works. Handy again if you ever want to sell your property. (N)

5. BUILDING, OVERVIEW

The Owner-Builder and the Code, Ken Kern, Ted Kogon, Rob Thallon, 1976. (Owner-Builder Publications, P.O. Box 817, North Fork, CA 93643) Read this book if you anticipate problems with the building code. There are lots of good case histories, interesting not only for the builders' experiences with building inspectors, but also as examples of low-cost owner-built homes. 2,6,N)
Shelter, edited by Lloyd Kahn, 1973. (Shelter Publications, P.O. Box 279, Bolinas, CA 94294) Jaki and I found this book freeing our preconceived notions on housing back in 1974. The examples of vernacular architecture from all over the world make these basic points time and again: The appropriate building material is all around you; the appropriate house style is dictated by the climate and landform. Good basic stick-framing information for small temporary shelters will be found here, but also hard-to-find info on sod and adobe construction, domes, bamboo, thatch, you name it. (1,2,6,9,N)

Shelter II, same people. Almost as good as *Shelter.* A little more up to date. Both *Shelter* books have excellent Bibliographies. (1,2,6,N)
Your Engineered House, Rex Roberts, 1964. (M. Evans and Co., 216 East 49th Street, New York, NY 10017) Rex says your house should be warm, dry, light, quiet, clean, useful and spacious; then he tells you how to design it that way, always with a view to low cost.

The book is 17 years old and could use an updating, but the basic premises are timeless. (1,2,6,N)

The Owner-Built Home, Ken Kern, 1972 and 1975. (Charles Scribner's Sons, Vreeland Avenue, Totowa, NJ 07512) A classic. Good overview. (2,6,9,N)
The Owner-Built Homestead, Ken Kern, 1974. (Scribner's) This one integrates shelter considerations with other living systems, especially food production. Recommended reading if you are using your mortgage-free home as a bridge to the self-reliant lifestyle. (4,12,13,N)
Building the House You Can Afford, Stu Campbell, 1979. (Garden Way, Charlotte, VT 05445) Step by step guide for building a traditional framed house on a standard foundation. Lots of valuable money-saving tips. (6)
In Harmony with Nature, Christian Bruyére and Robert Inwood, 1975. (Sterling Publishing Co., Inc., Two Park Avenue, New York, NY 10016) A constructional look at the kinds of houses featured in Boericke and Shapiro's books (see Category Two: Inspiration.) The accent is on log construction, roof and floor framing. Beautiful, clear illustrations. (2,6,8,9,N)
Country Comforts, Christian Bruyére and Robert Inwood, 1976. (Sterling) A good sequel to *In Harmony with Nature,* accenting the smaller outbuildings that are a part of the homesteading lifestyle: greenhouses, animal shelters, root cellars and the like. Same fine quality. (2,6,8,9,12,N)
New Shelter, Emmaus, PA 18049. An up-to-date magazine on all kinds of building, accenting energy-efficiency. Their February, 1981, issue, titled "How to Afford a New House" is particularly pertinent to readers seeking the mortgage-free home.

6. TIMBER FRAMING

The Timber Framing Book, Stewart Elliott and Eugenie Wallas, 1977. (Housesmiths Press, P.O. Box 157, Kittery Point, Maine 03905) Heavy duty post and beam framing techniques. How to do it the *right* way.
Woodframe Houses: Construction and Maintenance, L.O. Anderson, 1976. (Sterling Publishing Co., Inc. Two Park Avenue, New York, NY 10016) This is an updated version of Anderson's classic, *How to Build a Wood-Frame House.* Frank and Elizabeth Brasacchio (Chapter 6) found it invaluable in the building of their wood frame house. (N)

7. EARTH SHELTERING

Underground Houses: How to Build a Low-Cost Home, Robert L. Roy, 1979. (Sterling Publishing Co., Inc., Two Park Avenue, New York, NY 10016) Step by step construction details, accenting surface-bonded walls and plank-and-beam roofing. I know because I built my own subterranean house, lived in it and wrote this book! I think you'll find this book satisfying and complete. (N)
The $50 and Up Underground House Book, Mike Oehler, 1978. (Mole Publishing Co., Rt. 1, Box 618, Bonners Ferry, ID 83805) Truly the bottom line in low-cost earth-sheltering. Mike and I do not agree on some of the structural and design

philosophy details, but I commend his book for its imaginative approach to the problem of shelter. The $50 model ($22 went towards a stove and stovepipe) should be considered as a truly temporary shelter, but there is a value in that, as we have seen. (1,2,N)

Earth Sheltered Housing Design, The Underground Space Center, 1979. (Van Nostrand Reinhold, 7625 Empire Drive, Florence, KY 41042) Considered the Bible by those in the field. A bit technical in parts, but as complete a work as can be found under one cover. Not a how-to book, but very strong on design considerations for underground houses. Watch for a new companion volume. (N)

The Underground House Book, Stu Campbell, 1980. (Garden Way, Charlotte, VT 05445) Good clear overview. Up-to-date, easy to read and understand, excellent illustrations. Backed by the technical expertise of architect Don Metz. (N)

Earth Shelter Digest, 479 Fort Road, St. Paul, MN 55102. Stays on top of the underground movement. Contact with product manufacturers. Inspiring case histories in each bi-monthly issue. (N)

8. LOGS

Building with Logs, B. Allan Mackie, 1974, 1979. (Log House Publishing Co., Ltd., P.O. Box 1205, Prince George, British Columbia, Canada, V2L 4V3) Written by Canada's master log builder, with the same attention to detail displayed in his houses. Allen runs his log building school the same way. (N)

9. MASONRY (stone, adobe, cordwood, surface bonding, masonry stoves)

Stone

Stone Masonry, Ken Kern, Steve Magers, Lou Penfield, 1977. (Charles Scribner's Sons, Vreeland Avenue, Totowa, NJ 7512) J. Baldwin calls this "the best most comprehensive book on building with stone."[38] I agree. (N)

Adobe. Building with homemade adobe bricks is an appropriate building technique found mostly in the southwest. I have a lot of respect for adobe construction, but no experience. I suggest that readers interested in adobe start their search for information on page 229 of *The Next Whole Earth Catalog.* Four different books and an adobe journal are reviewed.

Cordwood

Cordwood Masonry Houses: A Practical Guide for the Owner-Builder, Robert L. Roy, 1980. (Sterling Publishing Co., Inc., Two Park Avenue, New York, NY 10016) Deals with all three styles of cordwood masonry: within a post and beam frame, within built-up corners, and as a load-bearing curved wall. Detailed.

Building the Cordwood Home, Jack Henstridge, 1977. (Indigenous Materials Housing Institute, Box H, Upper Gagetown, New Brunswick, Canada, E0G 3E0) A homemade book about a homemade house. Inspiring. Deals primarily with curved-wall construction. (2)

Surface bonding

Adhesive Bonding with Wood (Sterling Publishing Co., Inc., Two Park Avenue, New York, NY 10016) Manufacturers of surface bonding cement (Conproco, Surewall, etc.) can also supply building information.

Masonry Stoves

Masonry Stove Plans, Basilio Lepuschenko, Alexandria Road, Richmond, ME 04357. (N,*)

Masonry Stove Guild Newsletter, Maine Wood Heat Co., R.F.D.#1, Box 38, Norridgewock, ME 04957. (N,*)

Masonry Stoves, Albert A. Rarden III, Country Journal, February 1978, page 40. Excellent article.
On Building Masonry Firestoves, J. Patrick Manley, Farmstead Magazine #34, Fall 1980, page 36. Another informative article.

10. RENOVATION

Renovation: The Complete Guide, Mike Litchfield (available Spring 1982, John Wiley and Sons, Inc., 605 3rd Avenue, New York, NY 10016) This big book should be the definitive work for a long time to come. Mike sees renovation more as a strategy for people to apply to their existing homes, although he acknowledges that there might still be a few genuine bargains to be found "out in the sticks" for those seeking the mortgage-free home. (*)

11. KIT HOUSES

Build From a Kit, David Bear, Rodale's *New Shelter,* February, 1981, page 42. Realistic appraisal of the kit house strategy. Good source list of manufacturers (16 domes, 32 panels, 2 post and beams, 23 panel, 36 log). Mr. Bear lists these guides as reference:
The Complete Guide to Factory-Made Houses, A.M. Watkins, 1980. (E.P. Dutton, Two Park Avenue, New York, NY 10016) (*)
The Guide to Manufactured Homes, The National Association of Home Manufacturers, 6521 Arlington Blvd., Falls Church, VA 22042. (*)

12. FOOD

Fruits and vegetables. Two good sources of gardening literature accenting up-to-date organic methods are: Rodale Press, 33 East Minor Street, Emmaus, PA 18049 and Garden Way Publishing, Charlotte, VT 05445. Write for their free catalogs. Rodale's *Organic Gardening* is *the* gardening magazine.
How to Grow More Vegetables, John Jeavons, 1974, 1979. (Ten Speed Press, 900 Modoc Street, Berkeley, CA 94707) Biodynamic/French intensive raised bed gardening, the best way to raise vegetables at home. (N)
Grafting Manual, Larry Southwick, 1979. (Garden Way Publishing, Charlotte, VT 05445) A piece of land with "scrub" apples may be more valuable than you (or the seller) might think. Grafting is something like magic, except that it's real. (*)
Bees and livestock. Again, I have little experience in these fields, although others on the hill keep chickens, bees and goats. If I were seeking out the information, I would start with *The Next Whole Earth Catalog.* There are two pages of beekeeping access information alone, as well as books and magazines on rabbits, goats, chickens, pigs, and dairy cows.
Farmstead Magazine, Box 111, Freedom, ME 04941. If you can have "nuts and bolts" farmsteading information, this is it. (5,13)
Countryside Magazine, 312 Portland Road, Waterloo, WI 53594. Strong on livestock.

13. ENERGY

Wind
Windpower For Your Home, George Sullivan, 1978. (Cornerstone Library, Inc., New York) Good overview, bibliography.
Harnessing the Wind for Home Energy, Dermot McGigan, 1978. (Garden Way Publishing, Charlotte, VT 05445) J. Baldwin's choice for finding out about wind

energy. (Read Baldwin's comments on windpower on pages 196 and 197 of *The Next Whole Earth Catalog* before buying a wind system.) (N,*)
Wind Power Digest, 54468 CR 31, Bristol, IN 46507. Quarterly.
Photovoltaics
Photovoltaics, Sunlight to Electricity in One Step, Paul D. Maycock and Edward N. Stirewalt, 1981. (Brick House Publishing Co., 34 Essex Street, Andover, MA 01810) (*)
Making and Using Electricity from the Sun, the staff of the Solarex Corporation, 1335 Piccard Drive, Rockville, MD 20850. (*)
Passive solar
Natural Solar Architecture, a Passive Primer, David Wright, 1978. (Van Nostrand Reinhold, 7625 Empire Drive, Florence, KY 41042) The clearest explanation of how passive solar systems work that I have found. Let the sun heat your house.
30 Energy Efficient Houses You Can Build, Alex Wade and Neal Ewenstein, 1977. (Rodale Press, 33 East Minor Street, Emmaus, PA 18049) A book of passive solar homes, most of them owner-built at low cost. (That's why I particularly like this one.) Excellent plans, elevationals and photographs.

There is a tremendous amount of solar literature appearing today. Look it over carefully. A lot of it is woefully repetitive. Two or three good books is probably all you need to read. Again, consult *The Next Whole Earth Catalog,* which has a pretty good sampling, although the two fine books just listed above are not reviewed in the *Catalog.*
Alternative Sources of Energy, Milaca, MN 56353. Each bi-monthly issue zeroes in on a particular energy subject: wind, conservation, passive solar, wood-burning, etc. Nuts and bolts information made easy to digest.
Home Energy Digest, 8009 34th Avenue South, Minneapolis, MN 55420. This quarterly has a variety of articles on all aspects of energy production and conservation. The Fall, 1980, issue contains a special guide to owner-builder schools. Back issues can often be found where wood-burning stoves are sold—or write to the magazine itself.
How to Be Your Own Power Company (formerly *The Wilderness Home Power System*), Jim Cullen, 1978, 1980. (Van Nostrand Reinhold Co., 7625 Empire Drive, Florence, KY 41042) Good 12-volt power guide. How to use your car to run your house. (N)

14. URBAN HOMESTEADING

The Integral Urban House, The Farallones Institute Staff, 1979. (Sierra Club Books, Box 3886, Rincon Annex, San Francisco, CA 94119) Somehow or other you manage to build or renovate a town house. Now you want to derive the ongoing savings in food and fuel. Follow this guide and you'll probably meet with greater success than most country folks. (1,2,12,13,N,*)

15. MISCELLANEOUS

Drive It Till It Drops, Joe Troise, 1980. (And Books, The Distributors, 702 South Michigan, South Bend, IN 46618) Definitely worth a look for those readers operating on a real tight grubstake. (1,N,*)
Country Journal, P.O. Box 2402, Boulder, CO 80321. A surprising source of information on owner-building and soft technology. For example, the March, 1979, issue had five articles on homesteaders who fail. (Good issue to read.) (1,2,4,5,9, 12,13)

SOURCE NOTES

Reprinted by permission:
1. Kern, Ken; Kogon, Ted; and Thallon, Rob, *The Owner-Builder and the Code,* Charles Scribner's Sons, Copyright © 1976 by Ken Kern, Ted Kogon, and Rob Thallon, p. 78.
2. Ibid, p. 60.
3. Roy, Robert L., *Cordwood Masonry Houses: A Practical Guide for the Owner-Builder,* Sterling, N.Y. (1980), p. 132.
4. Ibid., p. 115.
5. Ibid., p. 135.
6. Borsodi, Ralph, *The Mother Earth News®* #26, "Plowboy Interview." Copyright © (1974/1981) by The Mother Earth News, Inc., P.O. Box 70, Hendersonville, N.C. 28791. One year subscription, $15.00. p. 11.
7. Kern, Ken; Kogon, Ted; and Thallon, Rob, op. cit., p. 18, quoted from: Sanderson, Richard L., *Codes and Code Administration,* Chicago: Building Officials Conference of America, Inc., 1969, p. 14.
8. Ibid., pp. 24–25, quoted from: National Commission on Urban Problems, *Building the American City,* Report of the National Commission of Urban Problems to the Congress and to the President of the United States, Washington, D.C.: U.S. Government Printing Office, 1968, p. 259.
9. Ibid., p. 25.
10. Rudofsky, Bernard, *The Prodigious Builders,* Harcourt Brace Jovanovich, New York, Copyright 1977 by Bernard Rudofsky, p. 271.
11. Ibid., p. 58.
12. Ibid., p. 62.
13. Ibid., p. 59.
14. Welsch, Roger L., *Shelter,* Shelter Publications, Bolinas, CA.
15. Boyles, Peg, *Country Journal,* January, 1981, "Solar Peasants," pp. 37–43.
16. Wells, Malcolm, *The CoEvolution Quarterly,* Fall, 1976, "Underground Architecture," p. 87.
17. Brand, Stewart, *The Next Whole Earth Catalog,* Random House, N.Y. (1980), p. 91.
18. Scher, Les, *Finding and Buying Your Place in the Country,* Macmillan, N.Y. (1974), p. 358.
19. Plattsburgh Press-Republican, Wed., January 7, 1981, "Energy Efficient Diet? Less Meat, More Grain and Vegetables," p. 9.
20. Brand, Stewart, op. cit., p. 292.
21. Easton, Bob, *Shelter,* Shelter Publications, Bolinas, CA (1973), p. 40.
22. Thoreau, Henry David, *Walden,* New American Library, Inc., N.Y. (1960) p. 38.
23. The Mother Earth News #67, "Bits and Pieces," p. 10.
24. Wing, Charles, *From The Ground Up,* Little, Brown and Company, Boston, Copyright © 1976, p. 69.
25. As quoted in *Shelter II,* Lloyd Kahn, editor, Shelter Publications, Bolinas, CA (1978) p. 74.
26. Pellman, Donald R., *Country Journal,* August, 1978, pp. 46–47.
27. Ibid., p. 44.
28. Oehler, Mike, *The $50 and Up Underground House Book,* Mole Publishing Co., Bonners Ferry,
29. ID (1978), p. 9.
    Campbell, Stu, *The Underground House Book,* Garden Way Publishing Company, Charlotte, VT,
30. Copyright 1980, p. 21.
    Earth Sheltered Housing Design, Van Nostrand Reinhold, Florence, KY (1979), p. 66.
31. Gaspers, Michael, *Shelter II,* op. cit., p. 85.
32. Gay, Larry, *Heating With Wood,* Garden Way Publishing Company, Charlotte, VT, Copyright 1974, p. 39.
33. Rudofsky, Bernard, op. cit., p. 276.
34. Kern, Ken; Kogon, Ted; and Thallon, Rob, op. cit., p. 84.
35. Wing, Charles, op. cit., p. 132.
36. Thoreau, Henry David, op. cit., p. 215.
37. Brand, Stewart, op. cit., p. 290.
38. Baldwin, J., op. cit., p. 231.

Index

add-on house strategy, 91, 92
area plan, 93
backfilling and landscaping, 85
basements, 96, 97, 98, 99
  capped, 99, 125
beekeeping, 36
Borsodi, Ralph, 17
braced posts, 82
building
  experience in, 53, 54
  full-time, 118
  part-time, 117
  rules for, 89-114,
  stages of, 121-122
building
  materials, recycled, 104
  regulations, 19, 23, 24
  site, 38
  strategies, 115, 116
  styles, 87, 88, 89
Building Officials Conference of America, 19
case histories, 123-145
casual ownership, 41
cedar
  posts, for shed, 57, 58, 59
  slabwood, 63
  tree, 74
"Centerville," maps of, 14, 18
code requirements, 23
construction stages, 87
cooperative land investment, 38, 39-41, 42
co-ownership, types of, 41
corporate ownership, 41
cordwood masonry
  infilling, 85
  in temporary structures, 65, 66, 68, 70
corporate ownership, 41
country land, 13-18, 19
deeds, types of, 42
design, 9, 10, 102
earth floor, 79
earth-shelter, 70, 112
Earthwood, 106-108, 110
economics of home building, 10, 11
Empiric Economic Laws, 11, 45
energy
  consumption, 107, 109, 110
  cost, 25
  sources, indigenous, 28, 29, 31
excavation, 87
floor plans, 99, 102, 103, 104
food
  consumption of, 50
  production of, 33-35, 36, 113, 114
framework
  post-and-beam, 70, 133
  wood floor, 60-61
French drains, 84
French intensive gardening, 33, 34
frost protection, 68, 70
fuel, 28, 51
Full Covenant and Warranty Deed, 42
Fuller, Thomas, 90

full grubstake, 43, 44-45, 46
gable, 56
gas well, 28
geodesic domes, 93
grubstakes, 43-52
  full, 43, 44-45, 46
  land, 46, 47-49, 50
home industry, 114-115
house construction
  rules for, 89, 90-115
  stages of, 120, 121-122
houses
  selection of, 87
  shapes of, 96
  types of, 87, 88-89
indigenous energy sources, 28, 29, 31
indigenous materials, availability of, 25, 26, 28
interim shelter costs, elimination of, 54, 55
joint tenancy, 41
kit houses, 120
labor barter, 115, 116
land
  acquisition, advantages of, 53, 54-55, 57
  grubstakes, 46, 47-49, 50
  values, influences on, 20-22
land, marginal, 36, 37, 38
land selection, 12-18, 19, 24-38
  economics of, 12-18, 19
  factors influencing, 24-36
  regulations covering, 19, 23, 24
livestock, 34, 35
living systems, 106-115, 116
log cabin
  interior of, 73, 74
  octagonal, 75, 143
Log End Cave, 116-117
Log End Cottage, 27
lumber milling, 62
maple syrup production, 36
marginal land, 36, 37, 38
masonry stove, 110, 111, 112
Miller, Rudolph, 19
mortgage payments, 10, 11
  methods of, 42, 47
National Commission on Urban Problems, 23
necessities and luxuries, 50, 51-52
on-cost cycle, 9
on line system, see utility interface system
overreaction syndrome, 90, 91
owner/builder as designer, 9, 10
owner/built homes,
  advantages of, 7-11
  case histories of, 123-145
photovoltaic cells, 31, 33
Pimental, David, 50
pipe plumb, 96
planning regulations, 19, 23, 24
Poisson, Leandre and Gretchen, 34
polygonal houses, 93, 94
posts
  braced, 82
  cedar, 59
probate courts, 49

property, definition of, 17
Quitclaim Deed, 42
rafters, 83, 98
raised beds, 33, 34, 35
recycled materials, 104
remote site system, 29
renovation, 118, 119, 200
restrictive planning, 39
reverse feed system, see utility interface system
ring beam, 70, 72
roofing nails, 62
round house, 94-95, 96, 97, 100-101
sauna
  cordwood, 70, 71
  floor plan for a, 76, 87
selection of houses, 81, 82, 87
selection of land, 12-18, 19, 24-38
sheds, 58-64
  gambrel-roofed, 67
shelters, temporary, 53-77
  advantages of, 53, 54-55, 57
  cordwood masonry in, 68, 69
  individual, 57, 58, 62-67
  underground, 70, 76, 77
site planning, 104, 105
slab, floating, 72
solar indexes chart, 32
solar power, 31, 33
solar radiation chart, 32
structures
  complex, 94
  easy-to-build, 103
Stump House, 69, 130-134
subterranean house, 117
sugar production, 36
surface bonding, 87
synchronous inverter, 31
temporary shelters
  advantages of, 53, 54-55, 57
  cordwood masonry technique in, 65-70
  types of, 57, 58-65
  underground, 70, 76, 77
tenancy-in-common, 41
tenants by entireties, 41
terms of purchase, 42
thermal mass, 109
Thoreau, David, 28, 62, 147
The Unincorporated Nonprofit Association, 41
United Nations Food and Agricultural Organization, 50
utility interface system, 31
vegetable production, 33, 34, 35
Walden Pond, 62
Warranty Deed, see Full Covenant and Warranty Deed
water power, 28
Welsch, Roger L., 26
window placement, 112, 113
wind power, 29
  averages of, 30
wind systems, 29, 31
Wing, Charles, 136
wood floor framing, 60-61
work parties, 116, 117
yurts, 93, 94, 95
zoning regulations, see planning regulations